POSITIVELY ORDERED SEMIGROUPS

PURE AND APPLIED MATHEMATICS

A Program of Monographs, Textbooks, and Lecture Notes

Contributions to *Lecture Notes in Pure and Applied Mathematics* are reproduced by direct photography of the author's typewritten manuscript. Potential authors are advised to submit preliminary manuscripts for review purposes. After acceptance, the author is responsible for preparing the final manuscript in camera-ready form, suitable for direct reproduction. Marcel Dekker, Inc. will furnish instructions to authors and special typing paper. Sample pages are reviewed and returned with our suggestions to assure quality control and the most attractive rendering of your manuscript. The publisher will also be happy to supervise and assist in all stages of the preparation of your camera-ready manuscript.

LECTURE NOTES
IN PURE AND APPLIED MATHEMATICS

1. *N. Jacobson,* Exceptional Lie Algebras
2. *L.-Å. Lindahl and F. Poulsen,* Thin Sets in Harmonic Analysis
3. *I. Satake,* Classification Theory of Semi-Simple Algebraic Groups
4. *F. Hirzebruch, W. D. Newmann, and S. S. Koh,* Differentiable Manifolds and Quadratic Forms
5. *I. Chavel,* Riemannian Symmetric Spaces of Rank One
6. *R. B. Burckel,* Characterization of C(X) Among Its Subalgebras
7. *B. R. McDonald, A. R. Magid, and K. C. Smith,* Ring Theory: Proceedings of the Oklahoma Conference
8. *Y.-T. Siu,* Techniques of Extension of Analytic Objects
9. *S. R. Caradus, W. E. Pfaffenberger, and B. Yood,* Calkin Algebras and Algebras of Operators on Banach Spaces
10. *E. O. Roxin, P.-T. Liu, and R. L. Sternberg,* Differential Games and Control Theory
11. *M. Orzech and C. Small,* The Brauer Group of Commutative Rings
12. *S. Thomeier,* Topology and Its Applications
13. *J. M. López and K. A. Ross,* Sidon Sets
14. *W. W. Comfort and S. Negrepontis,* Continuous Pseudometrics
15. *K. McKennon and J. M. Robertson,* Locally Convex Spaces
16. *M. Carmeli and S. Malin,* Representations of the Rotation and Lorentz Groups: An Introduction
17. *G. B. Seligman,* Rational Methods in Lie Algebras
18. *D. G. de Figueiredo,* Functional Analysis: Proceedings of the Brazilian Mathematical Society Symposium
19. *L. Cesari, R. Kannan, and J. D. Schuur,* Nonlinear Functional Analysis and Differential Equations: Proceedings of the Michigan State University Conference
20. *J. J. Schäffer,* Geometry of Spheres in Normed Spaces
21. *K. Yano and M. Kon,* Anti-Invariant Submanifolds
22. *W. V. Vasconcelos,* The Rings of Dimension Two
23. *R. E. Chandler,* Hausdorff Compactifications
24. *S. P. Franklin and B. V. S. Thomas,* Topology: Proceedings of the Memphis State University Conference
25. *S. K. Jain,* Ring Theory: Proceedings of the Ohio University Conference
26. *B. R. McDonald and R. A. Morris,* Ring Theory II: Proceedings of the Second Oklahoma Conference
27. *R. B. Mura and A. Rhemtulla,* Orderable Groups
28. *J. R. Graef,* Stability of Dynamical Systems: Theory and Applications
29. *H.-C. Wang,* Homogeneous Banach Algebras
30. *E. O. Roxin, P.-T. Liu, and R. L. Sternberg,* Differential Games and Control Theory II
31. *R. D. Porter,* Introduction to Fibre Bundles
32. *M. Altman,* Contractors and Contractor Directions Theory and Applications
33. *J. S. Golan,* Decomposition and Dimension in Module Categories
34. *G. Fairweather,* Finite Element Galerkin Methods for Differential Equations
35. *J. D. Sally,* Numbers of Generators of Ideals in Local Rings
36. *S. S. Miller,* Complex Analysis: Proceedings of the S.U.N.Y. Brockport Conference
37. *R. Gordon,* Representation Theory of Algebras: Proceedings of the Philadelphia Conference
38. *M. Goto and F. D. Grosshans,* Semisimple Lie Algebras
39. *A. I. Arruda, N. C. A. da Costa, and R. Chuaqui,* Mathematical Logic: Proceedings of the First Brazilian Conference
40. *F. Van Oystaeyen,* Ring Theory: Proceedings of the 1977 Antwerp Conference

41. *F. Van Oystaeyen and A. Verschoren,* Reflectors and Localization: Application to Sheaf Theory
42. *M. Satyanarayana,* Positively Ordered Semigroups

Other Volumes in Preparation

POSITIVELY ORDERED SEMIGROUPS

M. Satyanarayana

Professor of Mathematics
Bowling Green State University
Bowling Green, Ohio

MARCEL DEKKER, INC. New York and Basel

Library of Congress Cataloging in Publication Data

Satyanarayana, M.   [Date]
Positively ordered semigroups.

(Lecture notes in pure and applied mathematics ; v. 42)
Bibliography: p.
Includes index.
1. Semigroups. I. Title.
QA171.S27   512'.2   79-10077
ISBN 0-8247-6810-8

MARCEL DEKKER, INC.
270 Madison Avenue, New York, New York 10016

Current printing (last digit):
10 9 8 7 6 5 4 3 2 1

PRINTED IN THE UNITED STATES OF AMERICA

TO

VENKATESWARA SWAMY

CONTENTS

PREFACE

The study of positively totally ordered semigroups drew the attention of Hölder in connection with the problem of determining when a totally ordered semigroup is embeddable in the additive semigroup of positive real numbers. In recent years, interest in this study has been evinced by several people (A. H. Clifford and L. Füchs, for example). The theory of positively totally ordered semigroups has also found applications in measurement theory. The recent survey article by Jabovich (1976) on the fully ordered semigroups has stimulated the author to present a systematic account of positively totally ordered semigroups in greater detail. Most of the material in these notes was presented to faculty and graduate students at Bowling Green State University in a seminar for nearly two quarters during 1976-1977. The latest findings of the author are also included in these notes. This is the first attempt of the author to bring the results in this area in a consolidated form. Suggestions of the readers will be highly appreciated. It is hoped that this work will benefit many working in this field and stimulate them to handle the unsolved problems, some of which are mentioned at the end of this monograph. Finally, it is a great pleasure to acknowledge the help of C. S. Nagore for the useful changes in the original manuscript.

M. Satyanarayana

CHAPTER 1

INTRODUCTION

The purpose of this chapter is to introduce some basic concepts and the machinery needed to explain the new concepts which occur in the later chapters. We mention only some important definitions and results, which are used in the sequel. For the detailed proofs of some results mentioned here, we refer the reader to [1] for semigroup theory and to [2] for ordered structures. We begin with concepts in the theory of semigroups. Throughout this chapter let S denote an arbitrary semigroup; 0 is called a zero of S if $0 \cdot s = s \cdot 0 = 0$ for every $s \in S$. S is said to contain an identity 1 if $1 \cdot s = s \cdot 1 = s$ for every $s \in S$. If S contains 1, call $S = S^1$. Otherwise S^1 denotes $S \cup \{1\}$. If e is an element in S and if $e = e^2$, then e is called an idempotent. An ordering can be introduced among idempotents. If e and f are idempotents, $e < f$ if $e = ef = fe$. An element x in S is nilpotent if $x^n = 0$ for some natural number n, provided S contains 0. An element x in S is a left (right) zero-divisor if $xy = 0$ ($yx = 0$) for some nonzero y in S. x is a right (left) cancellable element in S if $yx = zx$ ($xy = xz$) for some y,z in S, then $y = z$. x is called cancellable if x is both right and left cancellative. x is a periodic element if $x^n = x^m$ for some natural numbers n and m.

A monoid is a semigroup with identity. If every element of S is nilpotent, then S is called a nilsemigroup. A nilsemigroup S is nilpotent if there exists a natural number n such that $x^n = 0$ for every $x \in S$. S is right (left) cancellative if every element of S is right (left) cancellative. S is cancellative if it is both right and left cancellative. S is periodic if every element of S is periodic. S is semisimple if $x \in SxSxS$ for every $x \in S$ or equivalently $A = A^2$ for every ideal A in S. S is called regular if, for every $a \in S$, there exists an $x \in S$ such that $a = axa$. S is a band if every element of S is an idempotent. A commutative band is called a semilattice. A chain is a semilattice in which the idempotents are linearly ordered under the ordering introduced above. If $S = S^2$, S is called globally idempotent. S is said to be periodic if every one of its elements is periodic.

NOTATION. '$\subset$' denotes inclusion and denotes proper inclusion. If A and B are subsets, $A \setminus B$ is the set of all elements in A which are not in B.

A nonempty subset T of S is a right (left) ideal if $s \in S$, $t \in T$ imply $ts \in T$ ($st \in T$); T is a two-sided ideal or simply an ideal if it is both a right and left ideal; a right or left or two-sided ideal is called proper if it is different from S. One-sided or two-sided ideal T is trivial if $S \setminus T$ is a singleton.

An ideal T is prime if $AB \subset T$ for some ideals A and B (or for some one-sided ideals of the same type), then $A \subset T$ or $B \subset T$; T is called completely prime if $ab \in T$ for some a and b in S, then $a \in T$ or $b \in T$; T is a maximal right (left) ideal if it is not contained in any proper right (left) ideal. R^*, L^*, M^*, P^* and Q^* denote the intersection of all maximal right ideals, maximal left ideals, maximal ideals, prime ideals and completely prime ideals respectively. Kernel of S is the intersection of all ideals. If S and T are semigroups, then a map $f : S \to T$ is called a homomorphism provided $f(ab) = f(a)f(b)$ for every a and b in S. With every ideal

A in S, we can define a canonical onto homomorphism from S onto S/A, where the elements of S/A are 0 (the image of A) and the other elements in S\A (S/A is called Rees-quotient semigroup); i.e., if $\bar{a},\bar{b} \in S/A$, then either a = b or $a,b \in A$.

Let $\{T_\alpha\}$ be a family of ideals in S such that $\bigcap T_\alpha = \emptyset$. Then S is said to be a subdirect product of S/T_α. This definition is a particular case of the general definition expressed in terms of congruences. R is a relation in S defined by: aRb iff $aS^1 = bS^1$; similarly L is a relation in S defined by: aLb iff $S^1a = S^1b$ and D is a relation defined by: aDb iff there exists an c such that aRc and cLb or vice versa. These relations partition S into disjoint subsets, which are called R-classes, L-classes and D-classes. A congruence relation ρ in S is an equivalence relation such that $xs\rho ys$ and $sx\rho sy$ for every $s \in S$, whenever $x\rho y$. The classes form a semigroup under the usual multiplication and is denoted by S/ρ. The association of the elements of S with its σ-equivalence classes is a homomorphism from S onto S/ρ. A congruence ρ is called a semilattice congruence if S/ρ is a semilattice.

S is called simple if S has no proper ideals. If S has no proper completely prime ideals, then S is called N-simple [1] and is also called ζ-indecomposable. The following is proved in [1].

FACT 1.1. *Every semigroup is a semilattice of N-simple semigroups; i.e., there exists a semilattice congruence, all of whose classes are N-simple semigroups.*

A right (left) ideal A in S is said to be finitely generated if $A = \bigcup_{i=1}^{n} x_i S^1$ $(A = \bigcup_{i=1}^{n} S^1 x_i)$. S is right (left) Noetherian if every right (left) ideal is finitely generated or equivalently every chain of right (left) ideals terminates at a finite stage. S is Archimedean if for every x and y in S there exists a natural number n such that $x^n \in SyS$. Clearly Archimedean semigroups are ζ-indecomposable but the converse need not be true.

A semigroup S is called a totally ordered semigroup, for short t.o. semigroup, if S is a totally ordered set under $\leq$ such that $a \leq b$ implies $ac \leq bc$ and $ca \leq cb$ for all $c \in S$. If '$a < b$' is not true, we write $a \geq b$. Assume that S is a t.o. semigroup here afterwords. For every element s in S such that $s \leq a$ ($a \leq s$) then a is called a maximal (minimal) element of S. An element a in S is a positive element if $x \leq ax$ and $x \leq xa$ for every s in S. If the inequality is strict, then a is called a strictly positive element. Similarly negative and strictly negative elements can be defined. S is called order-Archimedean (o-Archimedean) if a and b are positive elements such that $a^n < b$ for every positive integer n then a is the identity of S, if it exists and also if a and b are negative elements such that $a^n > b$ for every positive integer n then a is the identity of S, if it exists.

A semigroup S is said to satisfy right (left) quotient condition if, for every pair of elements a and b in S, there exist x and y in S such that $ax = by$ ($xa = yb$). Now, as mentioned in [2], we have

FACT 1.2. *Let S be a cancellative t.o. semigroup satisfying right quotient condition. Then S can be embedded in a t.o. group G of quotients* $g = ab^{-1}$, *where* $a,b \in S$. *If e is the identity of G, set* $g = ab^{-1} > e$ *iff* $a > b$. *G is then a t.o. group.*

A convex subset A of a t.o. semigroup S is a set in which $a < b < c$ with $a,c \in A$ imply $b \in A$. A congruence σ on a t.o. semigroup S is called convex if every σ-congruence class is a convex subset of S. If σ is a convex congruence on a t.o. semigroup S, then S/σ is a t.o. semigroup by prescribing:

$$\bar{a} \geq \bar{b},\ \bar{a},\bar{b} \in S/\sigma \text{ iff } a \geq b, \text{ where } a \in \bar{a},\ b \in \bar{b}$$

If S and T are t.o. semigroups, then f is called an order homomorphism (o-homomorphism) if f is a mapping such that

$f(ab) = f(a)f(b)$ for every a and b in S and $f(a) \leq f(b)$ whenever $a \leq b$. Order-isomorphism (o-isomorphism) is one to one onto o-homomorphism. If σ is a convex congruence on a t.o. semigroup S, then the canonical mapping $S \twoheadrightarrow S/\sigma$ is an o-homomorphism. In particular if A is a convex ideal in a t.o. semigroup S, then the natural map $S \twoheadrightarrow S/A$ is an o-homomorphism.

FACT 1.3. *If S is a t.o. semigroup, then the set E of all idempotents is S, if nonempty, is a subsemigroup of S.*

PROOF. Let $e,f \in E$. For definiteness let $e \leq f$. Then $e \leq ef$ and $fe \leq f$. This implies $ef = e^3f = e^2(ef) \leq efef$ and $efef = e(fe)f \leq ef^2 = ef$. Thus $ef = (ef)^2$. Hence $ef \in E$ and so E is a subsemigroup.

FACT 1.4. *If a t.o. semigroup contains positive and negative elements, then S contains idempotents.*

PROOF. Let x and y be positive and negative elements respectively. Then $xy \geq x$ and $xy \leq y$, which implies $x \leq y$. Since $x \leq x^3$, $xy \leq x^3y$ and $x \leq y$ implies $x^2 \leq xy$ and $yx \leq y^2$. Also $y^3 \leq y$. Hence $xy \leq x^3y = x^2(xy) \leq xyxy = x(yx)y \leq xy^3 \leq xy$. Thus xy is an idempotent.

CHAPTER 2

POSITIVELY TOTALLY ORDERED SEMIGROUPS

In this chapter we study the properties and structure of positively t.o. semigroups. A t.o. semigroup S is positively ordered if, for all a,b in S, $ab \geq a$ and $ab \geq b$. The first decomposition theorem of positively t.o. semigroups is due to Clifford and Klein-Barmen. They proved that every positively t.o. semigroup is an ordinal sum of ordinally irreducible semigroups. But it is a hopeless task to study the structure of ordinally irreducible semigroups since positively t.o. semigroups which are cancellative but do not contain identity are always ordinally irreducible and so, much cannot be said. Saito provided another decomposition for arbitrary t.o. semigroups [3]. According to Saito, two elements a and b in a t.o. semigroup are strongly o-Archimedean equivalent or said to belong to the same A-class, if there exist natural numbers p,q,r,s such that $a^p < b^q$ and $b^r < a^s$. The membership of the same A-class is an equivalence relation. If a t.o. semigroup contains only one A-class, then we say that S is strongly o-Archimedean. Saito proved that every t.o. semigroup is a pair-wise disjoint union of convex strongly o-Archimedean subsemigroups. In this representation, we do not know how to multiply the elements of two different strongly o-Archimedean classes. There exist only partial results in this direction [4].

However we provide an interesting decomposition theorem; every positively t.o. semigroup is a semilattice of o-Archimedean subsemigroups. This is a better theorem than Saito's in positively ordered case since the multiplication of different components is known. We observe also that positively o-Archimedean t.o. semigroups are nilsemigroups or subdirect product of nilsemigroups.

PROPOSITION 2.1. *For a positively t.o. semigroup S the following are true:*

(i) *x is a maximal element of S iff x is a zero of S;*

(ii) *If x is a minimal element, then $x = x^2$ or $S\backslash x$ is a maximal ideal. If $S = S^2$, then x is an idempotent if x is a minimal element;*

(iii) *If S contains identity 1, then 1 is the minimal element and $xy = 1$ for some $x,y \in S$ implies $x = y = 1$;*

(iv) *If $|S| > 1$, then S contains proper convex ideals. Every maximal one-sided and two-sided ideal is trivial;*

(v) *If $|S| > 1$, and if S contains an idempotent which is not a maximal element of S or if S contains two distinct idempotents, then S contains proper convex completely prime ideals;*

(vi) *If $x \leq e$ and $e^2 = e$, then $xe = ex = e$. The set of idempotents forms a chain under the dual ordering in semigroup-theoretic sense;*

(vii) *If $ab \leq ba$, then $a^nb^n \leq (ab)^n \leq (ba)^n \leq b^na^n$ and $a^nb^n \leq b^na^n$ for all natural numbers n and m.*

<u>*PROOF*</u>. (i) If x is a maximal element of S, then for every a in S, $ax \leq x$ and $xa \leq x$. The reverse inequality also holds since S is positively ordered and hence x is a zero. The other part is evident.

(ii) Assume that $S\backslash x$ is not an ideal. Then there exist $y \in S\backslash x$ and $z \in S$ such that $yz = x$, which implies by positive order that $x \geq y$ and $x \geq z$. Thus, by minimality of x, $x = y = z$

and hence $x = x^2$. Clearly if $S\backslash x$ is an ideal, it should be a maximal ideal. The second part is evident since the minimal element cannot be a product ot two elements unless it is an idempotent.

(iii) Since for every x in S, $x = x\cdot 1 \geq 1$, 1 is the minimal element. If $xy = 1$, then by positive order, $1 \geq x$ and hence $x = 1$ from the above. Therefore $y = 1$.

(iv) An easy verification shows that $T_a = \{x: x > a\}$ is a convex ideal. If $T_a = \emptyset$ for every $a \in S$, then $x \leq y \leq x$ for $x,y \in S$, which contradicts that $|S| > 1$. Hence at least one of $T_a \neq \emptyset$ and therefore S has proper convex ideals. Suppose that M is a maximal ideal and $S\backslash M$ contains two distinct elements x and y. Then $S = M \cup S^1xS^1 = M \cup S^1yS^1$, which implies that $x \in S^1yS^1$ and $y \in S^1xS^1$ and hence by positive order $x \geq y$ and $y \geq x$. Thus M is trivial. Similarly we can prove that maximal one-sided ideals are trivial.

(v) It is easy to verify that $T_e = \{x: x > e\}$ is a convex completely prime ideal, if e is an idempotent. If e is not a maximal element, then $T_e \neq \emptyset$. Thus S contains proper convex completely prime ideals. Let S contain two distinct idempotents e and f. One of T_e or T_f is non-empty since otherwise e and f are maximal elements and so $e = f$. Hence the conclusion is evident as above.

(vi) If $x < e$, then $xe \leq e$ and hence $xe = e$. Similarly $ex = e$. Thus, if e and f are two distinct idempotents then $ef = fe = e$ or $ef = fe = f$ according as $f < e$ or $e < f$. This implies that the set of all idempotents form a chain under the dual-ordering in semigroup-theoretic sense.

(vii) We shall prove the inequalities by induction. Here we shall indicate the proof for the first inequality and leave the proof of the second to the reader. For $n = 1$, the first inequality is true. Assume that this is true for $n - 1$. Then

$$a^nb^n = a(a^{n-1}b^{n-1})b \leq a(ba)^{n-1}b = (ab)^n$$

Similarly $(ba)^n \leq b^na^n$.

The kernel of a positively t.o. semigroup S, if exists, is zero since, if x and y are the elements in the kernel, then $S^1yS^1 = S^1xS^1$, which implies x = y. So the kernel contains a single element x and hence for every y in S we must have xy = x = yx and thus x is a zero. The same type of argument yields that every R-class, L-class and D-class contains a single element. By virtue of (iii) of 2.1, we can as well assure that positively t.o. semigroups do not contain identity.

An element x in a positively t.o. semigroup is said to be o-Archimedean, if for any y in S, there exists a natural number n such that $x^n \geq y$. By positive order we are assuming $x \neq 1$. Clearly a positively t.o. semigroup is o-Archimedean if every one of its is o-Archimedean. The following are some examples of o-Archimedean positively t.o. semigroups:

(i) The additive semigroup of positive real numbers with the usual order;

(ii) The set of all positive real numbers greater than 1 under multiplication and with usual order;

(iii) The set of all real numbers in (0,1] with the usual order and with the multiplication defined by:

$$a \circ b = \text{minimum of } a + b \text{ and } 1$$

Let S = {e,f,ef} where $e^2 = e$; $f^2 = f$ and ef = fe = f. Define the order by: e < f. This is positively t.o. semigroup which is not o-Archimedean.

PROPOSITION 2.2. *Let S be an o-Archimedean positively t.o. semigroup. Then the following are true:*

(i) If S does not contain identity, and if $x \in xS$ or $x \in Sx$ or $x \in SxS$, then x is an idempotent;

(ii) If $x \neq 1$ (identity), then x is an idempotent iff x is a maximal element or equivalently x is a zero;

(iii) *For every* $a \neq 1$ *in* S, $\bigcap_{n=1}^{\infty} a^n S = \bigcap_{n=1}^{\infty} Sa^n = \bigcap_{n=1}^{\infty} Sa^n S = 0$ *or* $\emptyset$ *according as* S *contains* 0 *or* S *does not contain* 0;

(iv) *If* S *does not contain identity,* $S \neq S^2$ *iff* S *contains maximal one-sided right, left, and two-sided ideals;*

(v) *If* $|S| > 1$, *then* S *contains* 1 *iff* $S = S^2$ *and* S *contains maximal one-sided and two-sided ideals;*

(vi) *If* $S = S^2$ *and if* S *is right or left Noetherian or* S *satisfies Noetherian condition on ideals, then* S *contains identity;*

(vii) *Every maximal one-sided ideal is trivial and two-sided;*

(viii) *If* S *contains* 1, *then* $R^* = M^* = L^* = S\backslash 1$. *If* S *does not contain* 1 *but contains maximal one-sided ideals, then* $R^* = L^* = M^* = S^2$;

(ix) *If* $a = ab$ *or* ba *for some* $a(\neq 1)$ *and* $b(\neq 1)$ *in* S, *then* S *contains a maximal element;*

(x) *If* $ab \leq ba$, *then* $(ba)^n \leq a^{n+1}b^n \leq b^n a^{n+1} \leq (ab)^{n+1}$ *for every natural number* n.

PROOF. (i) Let $x \in xS$. Then $x = xy$ for some y in S and so $x = xy^n$ for every natural number n. Hence $y^n \leq x$ for all n. This implies that $y = 1$ or $y^n = x$ for some n by o-Archimedean property. Hence $x = xy^n = x^2$. Similar proof can be given in the other two cases.

(ii) If $x \neq 1$ and if x is an idempotent, then the fact that $x < y$ for some $y \in S$ implies $x^n < y$ for every natural number n, which is impossible by o-Archimedean property. Thus x is a maximal element. Then by (i) of 2.1, x is a zero. The converse is evident.

(iii) Let $x \in \bigcap_{n=1}^{\infty} a^n S$. Then $x = a^n s_n$ for every natural number n and for some $s_n \in S$ and so $x \geq a^n$ by positive order. This implies $x = a^n$ for some n and so $x = xs_n$. Then by (i) x is an idempotent. So by (ii), if S contains 0, $x = 0$ and if S does not contain 0, x cannot exist. The other two cases can be wroked out in similar fashion.

(iv) Let $x \in S\backslash S^2$. Then $S\backslash x$ is a maximal one-sided ideal and also maximal ideal. To prove the converse it is sufficient to assume the existence of the maximal type of either two-sided or one-sided ideals. If M is a maximal right ideal, there exists an $x \in S\backslash M$, so that $S = M \cup x \cup xS$. Suppose $S = S^2$. Then $S = S^2 = MS \cup xS = M \cup xS$ and so $x \in xS$. Hence by (i) x is an idempotent and so is a zero by (ii). Hence $x = 0 \in M$, which is a contradiction.

(v) If S contains 1 and $|S| > 1$, then $S = S^2$ and also $S\backslash 1$ is the maximal one-sided and two-sided ideal by virtue of (iii) of 2.1. By (iv), the converse is evident.

(vi) This follows immediately from (v) since the Noetherian condition on one-sided or two-sided ideals implies the existence of maximal one-sided or two-sided ideals respectively provided that S is not right simple or left simple or simple. If S is one of the three types, namely, right simple, left simple or simple, then for every x in S we have $x \in xS$ or Sx or $x \in SxS$ respectively. So if $x \neq 1$, $x = 0$ by (i) and (ii). Thus $S = \{1\}$ or $S = \{0,1\}$ and hence S has an identity.

(vii) Triviality is evident from (iv) of 2.1. Suppose that M is a maximal right ideal which is not two-sided. Then there exists an $s \in S$ and $m \in M$ such that $sm \notin M$. In fact $s \notin M$ and so $S = M \cup s \cup sS = M \cup sm \cup smS$. Thus $s \in smS^1$. If S does not contain 1, then s is a zero by (i) and (ii) and so $s = 0 \in M$, which is a contradiction. If S contains 1, then $S\backslash 1 = M$ and so $sm = 1$. Hence $m = 1$ by (iii) of 2.1, which is again a contradiction. Similarly we can prove that every maximal left ideal is two-sided.

(viii) If S contains 1, then $S\backslash 1$ is the unique maximal two-sided and one-sided ideal by (iii) of 2.1. So $R^* = M^* = L^* = S\backslash 1$. Suppose that S does not contain identity. If $x \in R^*\backslash S^2$, then $S\backslash x$ is a maximal right ideal and so $x \in S\backslash x$. Therefore $R^* \subset S^2$. Let now $x \in S^2\backslash R^*$. Then there exists a maximal right ideal R not containing x. Thus $S = R \cup xS^1$ and $S^2 = RS \cup xS \subset R \cup xS$. Then $x \in xS$, which implies by (i) and (ii) that $x = 0 \in R^*$, which is a contradiction. Thus $R^* = S^2$. Since maximal one-sided ideals are trivial and two-sided, existence of one maximal type implies the existence of other types; L^* and M^* do exist and $L^* = M^* = S^2$ by similar reasoning.

(ix) Assume that $a = ab$. Let there exist an y in S such that $y > a$. We can choose a natural number n such that $b^n \geq y$. Therefore $a = ab^n \geq ay \geq y > a$, a contradiction. Similarly we can prove that a is a maximal element in the other case.

(x) If a is the identity, these inequalities are evident. Suppose that a is not the identity. Then by o-Archimedean property $a^{k-1} \leq b^n \leq a^k$ for some natural number k. Then by (vii) of 2.1, we have

$$(ba)^n \leq b^n a^n \leq a^{k+n} = a^{n+1} a^{k-1} \leq a^{n+1} b^n \leq b^n a^{n+1}$$

An analogous proof can be given for the last part.

We now describe all semisimple semigroups which are positively totally ordered.

THEOREM 2.3. *Let S be a semisimple semigroup. Then S is a positively t.o. semigroup iff S is a chain in which $e < f$, $e,f \in S$ if $ef = fe = f$.*

PROOF. Let S be a positively t.o. semigroup. If $a \in S$, then $a \in SaSaS$ and so $a = a^2$ by positive order. Then by (vi) of 2.1, S is a chain. The converse is evident.

COROLLARY 2.4. *The only simple semigroups which can be positively totally ordered are one-element semigroups.*

Another class of positively totally ordered semigroups which can be handled easily are o-Archimedean semigroups which contain maximal or minimal elements. These are described below:

THEOREM 2.5. *Let S be an o-Archimedean positively t.o. semigroup (possibly adjoined with 1). Then S is a nilsemigroup if S contains an idempotent ($\neq 1$) or S contains a maximal element or S contains elements $a(\neq 1)$ and $x(\neq 1)$ such that $a = ax$ or S contains 0. Conversely if S is a t.o. nilsemigroup with 0 as a maximal element (possibly adjoined with 1) then S is an o-Archimedean positively t.o. semigroup.*

PROOF. By virtue of (i) and (ii) of 2.1, the first two conditions on o-Archimedean positively t.o. semigroup S implies that S contains 0. Then for every $a(\neq 1)$ in S we have $a^n \geq 0$ for some n and hence $a^n = 0$, since 0 is the maximal element. Thus S is a nil-semigroup (possibly adjoined with 1). Conversely let S be a t.o. nilsemigroup with 0 as a maximal element. Suppose $ab < a$ for some a and b in S. Then $a \neq 0$ (and $b \neq 1$). By hypothesis $b^n = 0$ for some n. Now $ab^2 \leq ab < a$ and so $ab^2 < a$. Continuing in this manner we have $0 = ab^n < a$, which is not true. Thus S is positively ordered. Suppose now $x^n < y$ for every n and $x \neq 1$. Since some power of x is 0 by hypothesis, $0 < y$, which is not true. Thus S is o-Archimedean.

DEFINITION. A t.o. semigroup S is called a cyclic order extension of an infinite subsemigroup $\{x,x^2,\ldots\}$ if for every y in S we have $x^n < y \leq x^{n+1}$ for some n.

THEOREM 2.6. *The following are equivalent for any t.o. semigroup S not containing identity:*

(i) S is an o-Archimedean t.o. semigroup containing a minimal element;

(ii) (a) S is either a cyclic order extension of $\{x,x^2,\ldots\}$, where x is a minimal element, or

(b) S is a nilpotent semigroup containing a minimal element x such that $x^r = 0$ for some natural number r, where 0 is a maximal element.

PROOF. (i) $\Rightarrow$ (ii). Let x be a minimal element such that all powers of x are distinct. Then for any $y \neq x$, $x^m \geq y$ for some m by o-Archimedean property and $x < y$. So we can choose an n such that $x^{n-1} < y \leq x^n$. Thus S is of the form stated in (ii)a. Suppose $x^n = x^m$. Then for some r, x^r is an idempotent, which is zero by (ii) of 2.2. Since $x \leq y$, we have $x^r \leq y^r \leq x^r$ and so $y^r = x^r$. Thus S is a nilpotent semigroup.

(ii) $\Rightarrow$ (i). Assume (ii)a. If possible let $ab < a$ for some a and b in S. By hypothesis there exist natural numbers n and m such that $x^n < a \leq x^{n+1}$ and $x^m < b \leq x^{m+1}$. These inequalities together with $ab < a$ simply absurd statement. Thus S is positively ordered. To show that S is o-Archimedean assume first that $y^n < x^m$ for every n and for some y in S. Since $x \leq y$, $x^{m+1} \leq y^{m+1}$. Thus $x^{m+1} \leq y^{m+1} < x^m < x^{m+1}$, which is absurd. Let now $y^n < z$ for every n and $z \neq x^m$ for any m. By hypothesis, $z \leq x^t$ for some natural number t and so $y^n < x^t$, which is proved to be not true. Hence S is o-Archimedean. Clearly (ii)b implies (i) by 2.5.

Recall that a semigroup S is finitely generated as a right ideal if $S = \bigcup_{i=1}^{n} x_i S$ for some x_i's in S. S is finitely generated if there exist a finite number of elements $x_1, x_2, \ldots, x_n$ in S such that every element is a product of these.

COROLLARY 2.7. *Let S be an o-Archimedean positively t.o. semigroup not containing identity and assume that S is finitely generated or S contains a finitely generated maximal right ideal. Then S is of the form described in 2.6.*

PROOF. By virtue of 2.6, it suffices to show that S has a minimal element in both cases. If $S = \bigcup_{i=1}^{n} x_i S^1$, then it can be verified easily that the minimal among x_i's is the minimal in S. In the second case if a maximal right ideal M is of the form

$\bigcup_{i=1}^{n} y_1S^1$, then for any $x \in S\backslash M$, we have $S = \bigcup_{i=1}^{n} y_iS^1 \cup xS^1$. Thus this case is reduced to the first one.

In the following we describe o-Archimedean positively t.o. semigroups.

THEOREM 2.8. *Let S be an o-Archimedean positively t.o. semigroup. If S contains 0, then S is a nilsemigroup, possibly adjoined with 1. If S does not contain 0, then S is a subdirect product of nilsemigroups (possibly adjoined with 1) and S\1 is o-homomorphic with a subsemigroup of additive group of real numbers with the usual order.*

PROOF. The characterization of S containing 0 is evident from 2.5. Suppose that S does not contain 0. By (iii) of 2.1, S can be treated as an o-Archimedean t.o. semigroup T adjoined with 1 if S contains 1. Evidently T does not contain 0 and does not contain idempotents by (ii) of 2.2. Then $\bigcap_{n=1}^{\infty} T^1a^nT^1 = \emptyset$ for any $a \in T$ by (iii) of 2.2. Set $T_a = \{x\colon x \in T \text{ and } a \notin T^1xT^1\}$. Clearly T_a is an ideal and $a \notin T_a$ by (iii) of 2.2. Hence $\bigcap_{a\in T} T_a = \emptyset$. So S is a subdirect product of T/T_a, $a \in T$. Let $x \in S$. Since $\bigcap_{i=1}^{\infty} T^1x^iT^1 = \emptyset$, $a \notin T^1x^iT^1$ for some integer i. So $x^i \in T_a$ and T/T_a is nil for all $a \in T$. The second part follows from a result of Kowalski [5], which we discuss in 4.11.

Having known the structure of o-Archimedean positively t.o. semigroups fairly well, we shall now describe the structure of arbitrary positively t.o. semigroups in terms of o-Archimedean positively t.o. semigroups. The following lemma is a crucial result.

LEMMA 2.9. *Let S be a positively or negatively t.o. semigroup. If S is ζ-indecomposable, then S is o-Archimedean. In addition if $|S| > 1$, then S does not contain identity.*

PROOF. Let S be positively ordered. Consider $S_x = \{y: y \in S \text{ and } y^n \geq x \text{ for some } n\}$. Since $x \in S_x$, $S_x \neq \emptyset$. Clearly S_x is an ideal. Let $ab \in S_x$ with $a,b \notin S_x$. Then $(ab)^n \geq x$ for some n. Assume for definiteness $a \geq b$. Then $a^2 \geq ab$. Therefore $x \leq (ab)^n \leq a^{2n} < x$ since $a^m < x$ for every m. This is absurd. Hence S_x is a non-empty completely prime ideal. This implies $S_x = S$ for every x since S is ζ-indecomposable. Thus S is o-Archimedean. For negatively ordered semigroups similar proof can be given by considering $S_x^* = \{y \in y \in S \text{ and } y^n \leq x \text{ for some } n\}$. If $|S| > 1$ and if S contains 1, then $S\backslash 1$ is a completely prime ideal because of (iii) of 2.1, which is a contradiction.

Combining 1.1 and 2.8, we have,

THEOREM 2.10. *If S is a positively t.o. semigroup, then S is a semilattice of o-Archimedean positively t.o. semigroups S_α, where S_α contains no identity if $|S_\alpha| > 1$.*

THEOREM 2.11. *Let S be positively t.o. semigroup. Then S is periodic iff S is a disjoint union of nilsemigroups.*

PROOF. Let S be periodic. By 2.10, $S = \cup S_\alpha$ where S_α is o-Archimedean and S_α contains no identity if $|S_\alpha| > 1$. If $|S_\alpha| = 1$, then S_α is clearly a nilsemigroup. Assume that $|S_\alpha| > 1$ and hence S_α has no identity. Let $x \in S_\alpha$. Since x is periodic, there exists a natural number n such that x^n is an idempotent. Since S_α is o-Archimedean without identity, x^n is a zero of S_α by (ii) of 2.2 and hence is a nilsemigroup by 2.5. The converse is evident.

In literature there exists another decomposition theorem given by Klein-Barmen and Clifford. For this we need the following concept.

DEFINITION. Let Λ be a t.o. set and $\{S_\lambda\}$, $\lambda \in \Lambda$ be a family of t.o. semigroups such that $S_\lambda \cap S_\mu = \emptyset$ if $\lambda \neq \mu$. Then a semigroup S is called an ordinal sum of t.o. semigroups $\{S_\lambda\}$ if $S = \bigcup_{\lambda \varepsilon \Lambda} S_\lambda$ and the ordering and multiplication in S are given by the following rules:

If $a \varepsilon S_\lambda$ and $b \varepsilon S_\mu$ and $\lambda < \mu$ then set $a < b$ and $ab = ba = b$;

If a and b belong to the same S_λ, assume that the multiplication and ordering are the same as originally given in S_λ.

It is evident that S is positively ordered iff every S_λ is positively ordered. A t.o. semigroup is said to be *ordinally irreducible* if it cannot be expressed as an ordinal sum of two (or more) of its subsemigroups.

PROPOSITION 2.12. *Let S be an o-Archimedean positively t.o. semigroup without identity. Then S is ordinally irreducible.*

PROOF. Suppose that S is an ordinal sum of ordinally irreducible semigroups S_λ, where the ordering is as described in the definition. Let $a \varepsilon S_\lambda$ and $b \varepsilon S_\mu$ and $\lambda < \mu$. Then $a < b$ and $ab = ba = b$. By o-Archimedean property there exists a positive integer n such that $a^n \geq b$. If $a^n = b$, then $b \varepsilon S_\lambda \cap S_\mu = \emptyset$. So $a^n > b$. Then again by the definition of order $a^n = a^n b = b$, which is a contradiction. So S is ordinally irreducible.

THEOREM 2.13. *Let S be a cancellative positively t.o. semigroup. Then S is ordinally irreducible iff S does not contain 1.*

PROOF. Let S be ordinally irreducible. Suppose that S contains 1. Since S is positively ordered, S has no invertible elements except 1 by (iii) of 2.1. So $S\backslash 1$ is a subsemigroup. Now $S = S_1 \cup S_2$, where $S_2 = S\backslash 1$ and $S_1 = \{1\}$. Take $\Lambda = \{1,2\}$. Since S is positively ordered $a \varepsilon S_1$, $b \varepsilon S_2$ imply $a = 1 < b$. Clearly $ab = ba = b$. Thus S is the ordinal sum of S_1 and S_2, which is a contradiction. Thus S does not contain 1. Conversely let S have no identity. Suppose that S is the ordinal sum of ordinally irreducible subsemigroups S_α, $\alpha \in \Lambda$. For $\alpha < \beta$, $\alpha,\beta \in \Lambda$ we have $ab = ba = b$ where $a \in S_\alpha$ and $b \in S_\beta$. Therefore $b = ab = ab^2$, which implies $b = b^2$ by cancellative condition. Hence $b = 1$ since S is right and left cancellative. This is a contradiction.

From the above theorem one can observe that ordinally irreducible positively t.o. semigroups do not contain identity.

PROPOSITION 2.14. *A positively t.o. semigroup S is ordinally irreducible if S satisfies any one of the following conditions:*

(i) S is right cancellative semigroup without idempotents;

(ii) $\bigcap_{n=1}^{\infty} a^n S = \emptyset$ *for every a in S.*

PROOF. Suppose $S = \bigcup S_\alpha$ with $\alpha \in \Lambda$, $|\Lambda| > 1$ as in the definition. If $a \in S_\lambda$ and $b \in S_\mu$ with $\lambda < \mu$, then $ab = ba = b$. Then $b \in \bigcap_{n=1}^{\infty} a^n S$. So (ii) is not true. If (i) holds, $ab = b$ implies $ab = a^2 b$, which implies $a = a^2$. This is not true. Hence S is ordinally irreducible.

THEOREM 2.15. *Every positively t.o. semigroup can be uniquely represented as an ordinal sum of a t.o. set of ordinally irreducible positively t.o. semigroups.*

PROOF. Let S be a positively t.o. semigroup. (L,U) is called a cut in S if $L \cap U = \emptyset$; $a \in L$, $x < a$ imply $x \in L$ and $b \in U$, $y > b$ imply $y \in U$. L and U are called lower and upper classes of the cut. (L,U) is called an α-cut if L is a subsemigroup of S, and $a \in L$ and $b \in U$ imply $ab = ba = b$. The α-cuts can be totally ordered by setting $(L_1,U_1) < (L_2,U_2)$ if L_1 is properly contained in L_2. An α-cut is called a β-cut if it has an immediate successor α-cut in this ordering. Let Λ denote the totally ordered set of all β-cuts of S. Corresponding to every $\lambda \in \Lambda$, define S_λ as the intersection of the upper class U_λ of the β-cut (L_λ, U_λ) with the lower class $L_\lambda{}^1$ of its immediate successor α-cut $(L_\lambda{}^1, U_\lambda{}^1)$. Since U_λ and $L_\lambda{}^1$ are convex subsemigroups, S_λ is also a convex subsemigroup. We claim that the S_λ are ordinally irreducible and S is their ordinal sum.

Suppose that S_λ is an ordinal sum of its subsemigroups $S_\lambda{}^1$ and $S_\lambda{}^2$. It can be verified easily that $(L_\lambda \cup S_\lambda{}^1, S_\lambda{}^2 \cup U_\lambda{}^1)$ is an α-cut between (L_λ, U_λ) and $(L_\lambda{}^1, U_\lambda{}^1)$, which contradicts that $(L_\lambda{}^1, U_\lambda{}^1)$ is the immediate successor (L_λ, U_λ). Hence S_λ is ordinally irreducible. $T = \bigcup_{\lambda \in \Lambda} S_\lambda$, is clearly an ordinal sum of S_λ's and is contained in S. We claim that S = T. It suffices now to prove that every $a \in S$ is in some S_λ. Corresponding to each $a \in S$, consider the union L_0 of all lower classes of α-cuts (L,U) with $a \notin L$ (including the void set) and the intersection L_1 of all L's with $a \in L$. Then $(L_0, S \backslash L_0)$ and $(L_1, S \backslash L_1)$ are α-cuts in S, and it can be proved easily that no α-cut lies between them. Therefore $(L_0, S \backslash L_0)$ is a β-cut, and if S_λ corresponds to this β-cut, then $a \in S_\lambda$.

To prove the uniqueness, assume that S has another representation of the ordinal sum of ordinally irreducible t.o. semigroups T_μ $(\mu \in M)$. It is easy to verify that $L_\delta = \bigcup_{\mu<\delta} T_\mu$, $U_\delta = \bigcup_{\mu \geq \delta} T_\mu$ define an α-cut (L_δ, U_δ) and $L_\delta{}^1 = \bigcup_{\mu \leq \delta} T_\mu$, $U_\delta{}^1 = \bigcup_{\mu > \delta} T_\mu$ define an α-cut $(L_\delta{}^1, U_\delta{}^1)$. Between these α-cuts there exists no other α-cut since every T_μ is ordinally irreducible. Hence (L_δ, U_δ) is a β-cut of S with $(L_\delta{}^1, U_\delta{}^1)$ as immediate successor α-cut. Consequently, for some $\lambda \in \Lambda$, $S_\lambda = U_\delta \cap L_\delta{}^1 = T_\delta$. Thus every S_λ occurs as a T_δ and conversely every T_δ occurs as a S_λ. Hence the representation is unique.

A semigroup S is *separative* if for any $x, y \in S$, $x^2 = xy$ and $y^2 = yx$ imply $x = y$, and $x^2 = yx$ and $y^2 = xy$ imply $x = y$. By 1.1 and the Corollary 6.5 of [1;52], every separative semigroup is a semilattice of N-classes, where each N-class is a ζ-indecomposable cancellative semigroup. Since ζ-indecomposable positively t.o.

semigroups are o-Archimedean by 2.9 and hence ordinally irreducible if they do not contain identity, each N-class of a separative positively t.o. semigroup is a cancellative ordinally irreducible positively t.o. semigroup if it contains more than one element, and otherwise also it is trivially the same. Thus every separative positively t.o. semigroup is a semilattice of cancellative ordinally irreducible semigroups. It is very hard to describe the general situation. However Conrad [6] provided the following result for a special case.

THEOREM 2.16. *Let S be a positively t.o. semigroup. Then S is an ordinal sum of cancellative ordinally irreducible t.o. semigroups iff S satisfies the condition:*

$ab = ac$ (or $ba = ca$) implies $b = c$ or $ab = a$ ($ba = a$)

PROOF. It is easy to verify that S satisfies the condition if S is an ordinal sum of cancellative ordinally irreducible t.o. semigroups. Conversely let S satisfy the above condition. We note first that whenever ab in S is an idempotent, then either a or b is an idempotent and ab is the maximum of a and b. With the help of this property, an equivalence relation ρ on S can be defined by: $a \rho b$ if $a = b$ or if a and b are not idempotents and minimum of ab and ba is greater than the maximum of a and b. It can be proved that the equivalence classes S_λ are cancellative subsemigroups which are ordinally irreducible. Since S_λ's are convex, we can define $S_\lambda < S_\mu$ and so for $a \in S_\lambda$, $b \in S_\mu$ $a < b$ holds. It can be verified that S is the ordinal sum of S_λ's.

Every right Noetherian semigroup is finitely generated as a right ideal. But right Noetherian semigroups need not be finitely generated. In the case of o-Archimedean positively t.o. semigroups we shall show below that these two concepts coincide without even satisfying Noetherian condition.

THEOREM 2.17. *If S is an o-Archimedean positively t.o. semigroup without identity and if S is finitely generated as a right (left) ideal, then S is finitely generated.*

<u>*PROOF.*</u> Let $S = \bigcup_{i=1}^{n} x_i S$ and assume that $x_1 < x_2 < \ldots < x_n$. Let z be an element of S which is not a product of x_i's. Then $z = x_i s$, so that $z \geq x_i$. Again by the choice of z, $s = x_j t$. Then $s \geq x_j \geq x_i$. Thus $z \geq {x_1}^2$ and hence by induction $z \geq {x_1}^n$ for all positive integers n. This implies $z = {x_1}^n$ for some n, by o-Archimedean property. This is a contradiction. Therefore S is finitely generated by x_i's. The second case can be treated similarly.

COROLLARY 2.18. *If S is right Noetherian or left Noetherian o-Archimedean positively t.o. semigroup without identity, then S is finitely generated.*

COROLLARY 2.19. *If an o-Archimedean positively t.o. semigroup S contains a finitely generated maximal right (left) ideal and if S does not contain identity, then S is finitely generated.*

<u>*PROOF.*</u> Let $\bigcup_{i=1}^{n} x_i S^1$ be a maximal right ideal. Then $S = \bigcup_{i=1}^{n} x_i S^1 \cup yS^1$, where $y \notin M$. Then by 2.17, the result follows. Similar proof can be given in the case when S contains a finitely generated maximal left ideal.

In 2.9 we have proved that ζ-indecomposable positively t.o. semigroups are o-Archimedean. But the converse is not true, which can be seen in the example of the multiplicative semigroup of natural numbers with the usual order. But we prove below that these concepts coincide if the positively t.o. semigroup is principally generated as a right or left ideal.

THEOREM 2.20. *The following conditions on a semigroup* S *with* $|S| > 1$ *are equivalent:*

(i) S *is an* ζ*-indecomposable positively t.o. semigroup and* $S = xS^1$, $x \in S$;

(ii) S *is an o-Archimedean positively t.o. semigroup and* $S = xS^1$, $x \in S$;

(iii) S *is an infinite cyclic semigroup generated by* x *and ordered by:* $x < x^2 < \dots$ *or* $S = x, x^2, \dots, x^n = 0$ *where* $x < x^2 < \dots < x^n = 0$.

PROOF. By 2.9, (i) $\Rightarrow$ (ii). Assume (ii). Then by 2.17, S is generated by x. If no two powers of x are the same, then S is an infinite cyclic semigroup with that prescribed order. If $x^n = x^m$ for some natural numbers n and m, then for some r, x^r is an idempotent, which is a zero and a maximal element by (ii) of 2.2 provided $x^r \neq 1$. If $x^r = 1$, then $x = 1$ and so $|S| = 1$ by (iii) of 2.1. Hence (iii) is evident. (iii) $\Rightarrow$ (i) is trivial.

It is interesting to enquire whether there exist positively t.o. semigroups, which are finitely generated as a right ideal but not o-Archimedean. Under some restriction it is possible to describe all these semigroups.

THEOREM 2.21. *Let* S *be a positively t.o. semigroup without identity and* $S = xS^1$ *for some* $x \in S$. *If* $\bigcap_{n=1}^{\infty} x^n S = yS^1$ *and* y *is an o-Archimedean element, then* S *is one of the following:*

(i) $S = \{x, x^2, \dots, x^n = y = 0\}$ *with* 0 *as a maximal element;*

(ii) Every element of S *is of the form* $y^r x^s$, $r, s \geq 0$;

$y = xy$ *and* $x \leq x^2 \leq \dots \leq y \leq yx \leq yx^2 \leq \dots \leq y^2 \leq y^2 x \leq \dots$. *The second case happens iff* S *is not o-Archimedean.*

PROOF. Since $S = xS^1$, every element of S is a power of x or is an element in $\bigcap_{n=1}^{\infty} x^nS = yS^1$. So if $y \neq x^n$ for any natural number n, then $y = xs$, where s is not a power of x and $y \geq s$. But $s \in \bigcap_{n=1}^{\infty} x^nS = yS^1$ and so $s \geq y$. Hence $s = y$ and thus $y = xy$. Therefore we have $y = x^m$ for some m or $y = xy$. Then $x^m \in x^{m+t}S$ for any $t \geq 1$ and so $x^m = x^{m+t}$ by positive order. Hence $y = x^m$ is an idempotent. We claim now that y is a zero of S. Since y is an o-Archimedean element for any $s \in S$ there exists a natural number n such that $y = y^n \geq ys \geq y$ and so $y = ys$. Similarly $y = sy$. Thus y is a zero of S and so $S = \{x, x^2, \ldots, x^m = y = 0\}$. Suppose now $y = xy$. Then every element of S is of the form y^rx^s, $r \geq 0$, $s \geq 0$. For, if z is an element which is not of this form, then $z = ys_1$, where s_1 is not of the above form. Thus $s_1 = ys_2$, where s_2 is not of the above form. Then $z = ys_1 = y^2s_2$, which implies $z \geq y$ and $z \geq y^2$. By induction we have $z \geq y^n$ for every natural number n. Then by o-Archimedean property of y, $y^n = z$ for some n, which contradicts our supposition. Thus the second characterization is evident.

If S is not o-Archimedean, S should be of type (ii) since nilsemigroups with zero as a maximal element are o-Archimedean. Since $x^n < y$ for every natural number n, semigroups of type (ii) are not o-Archimedean.

In 2.20, we have proved that any o-Archimedean positively t.o. semigroup which is principally generated as a right ideal is an infinite cyclic semigroup if it does not contain a maximal element. However, for positively t.o. semigroups which are not o-Archimedean but principally generated as a right ideal, we can exhibit an o-homomorphism onto infinite cyclic semigroups possibly adjoined with 0 in certain cases. A weaker form of this result but with a stronger hypothesis of o-Archimedean condition is obtained by Kowalsky and Hion. They have shown that cancellative o-Archimedean positively t.o. semigroups and o-Archimedean positively t.o. semigroups without maximal elements can be homomorphically mapped into a subsemigroup of the additive semigroup of non-negative real numbers. We shall discuss these results in Chapter 4.

THEOREM 2.22. *Let S be a non-globally idempotent positively t.o. semigroup. If $S = xS^1$ for some $x \in S$ and if S is either left cancellative or right cancellative or S does not contain idempotents, then S is o-isomorphic with an infinite cyclic semigroup or there exists an o-homomorphism of S onto an infinite cyclic semigroup adjoined with 0.*

PROOF. It suffices to consider the case when S has no idempotents. For, let S be left cancellative or right cancellative. If S contains an idempotent e, then for any $x \in S$, we have $ex = e^2x$ and $xe = xe^2$ and so $x = ex \in S^2$ and $x = xe \in S^2$ according as S is left or right cancellative, which contradicts the hypothesis $S \neq S^2$. Since $S = xS^1$, we have $S^{r+1} = x^rS$ for every natural number r and $x^r \notin \bigcap_{n=1}^{\infty} S^n = S^{\omega}$ for any natural number r since otherwise $x^r \in S^{r+2} = x^{r+1}S$, which implies $x^r = x^{r+1}$ is an idempotent, which is not true. Suppose $S^{\omega} = \emptyset$. Then for any $y \in S$, $y \in S^n \backslash S^{n+1}$ for some n depending on y. Therefore $y = x^{n+1}s$ where $s = x$ or xt. Since $y \notin S^{n+1}$, we must have $y = x^n$. Since S has no idempotents, no two different powers of x coincide. Thus S is an infinite cyclic semigroup. Let now $S^{\omega} = \emptyset$. Then $S \neq S^{\omega}$ since $S \neq S^2$. We claim now that S^{ω} is a convex ideal. Suppose $a < b < c$ and $a, c \in S^{\omega}$. If $b \notin S^{\omega}$, $b = x^r$ for some natural number r, as above. But $a \in S^{r+2} = x^{r+1}S$. Therefore $x^{r+1} \leq a < b = x^r$, which is not true by positive order. Hence $b \in S^{\omega}$ and thus S^{ω} is a convex ideal. Then $\bar{S} = S/S^{\omega}$ is a t.o. semigroup by defining $\bar{a} < \bar{b}$ with $\bar{a}$ and $\bar{b}$ in $\bar{S}$ iff there exist an $a \in \bar{a}$ and $b \in \bar{b}$ such that $a < b$. Evidently $\bar{S}$ is a positively t.o. semigroup with zero $(\bar{0})$ and $\bar{S} = \bar{x} \cup \bar{x}\bar{S}$, where $\bar{x}$ is the image of x under the canonical mapping $S \twoheadrightarrow \bar{S}$. It can easily be verified that $\bar{S}^{\omega} = \bar{0}$. As above if $\bar{0} \neq \bar{a} \in \bar{S}$, then $\bar{a} = \bar{x}^r$ for some natural number r.

Also $\bar{x}^r \neq \bar{x}^{r+1}$ for any natural number r since $\bar{x}^r = \bar{x}^{r+1}$ implies that $x^r = x^{r+1}$ is an idempotent or $x^r \in S^\omega$. Both are impossible. Thus $\bar{S}$ is an infinite cyclic semigroup adjoined with $\bar{0}$ and the canonical homomorphism provides the necessary o-homomorphism.

In 2.9 we have noted that a positively t.o. semigroup is o-Archimedean if it has no proper completely prime ideals. In fact we will establish the converse.

THEOREM 2.23. *Let S be a positively t.o. semigroup. Then S is o-Archimedean iff S contains no proper convex completely prime ideals.*

PROOF. Let S be o-Archimedean. Suppose that P is a convex completely prime ideal and $P \neq S$. Then, if $x > a$ for some $x \in S\backslash P$ and $a \in P$, $a < x \leq ax$, which implies $x \in P$ by convexity of P. Hence $a > x$ whenever $a \in P$ and $x \in S\backslash P$. But by o-Archimedean property of S, $x^n \geq a$ for some natural number n. But $x^n \in S\backslash P$, since $S\backslash P$ is a subsemigroup, which contradicts the above assertion. The converse is evident by noting that the sets S_x occurring in the proof of 2.9 are convex completely prime ideals.

Since the structure of o-Archimedean positively t.o. semigroups are known in 2.8, the questions which now arise are when a positively t.o. semigroup contains proper completely prime ideals and in that case what is their structure. The following are some sufficient conditions for the existence of completely prime ideals. One may note that o-Archimedean positively t.o. semigroups might contain proper completely prime ideals as the example of the multiplicative semigroup of natural numbers greater than 2 with the usual order, indicates.

PROPOSITION 2.24. *In a positively t.o. semigroup S, the set A of all o-Archimedean elements in S is a completely prime ideal, if nonempty.*

PROOF. Let $a \in A$. Then for every $b \in S$, then there exists a natural number depending on b such that $a^n \geq b$. Then for $s \in S$, $(as)^n \geq a^n \geq b$ and $(sa)^n \geq a^n \geq b$ and so as and sa belong to A. Thus A is an ideal. A is completely prime. For, let $ab \in A$ with $a,b \notin A$. Assume $a \leq b$ for definiteness. Now $b^n < c$ for some c and for every natural number n, so that $(ab)^n \leq b^{2n} < c$ for every n. This contradicts that $ab \in A$.

PROPOSITION 2.25. *Let S be a positively t.o. semigroup in which every right ideal is two-sided. Then* $T = \{x: \bigcap_{n=1}^{\infty} x^n S = \emptyset\}$ *is an ideal, if nonempty. Furthermore if S is commutative, then T is completely prime.*

PROOF. Let $x \in T$. Then $\bigcap_{n=1}^{\infty} x^n S = \emptyset$. If $s \in S$, since $x \in xS^1$, which is two-sided, $sx \in xS^1$ and so $(xs)^n S \subseteq x^n S$. Thus, if $y \in \bigcap_{n=1}^{\infty} (xs)^n S$, then $y \in \bigcap_{n=1}^{\infty} x^n S = \emptyset$. Hence $\bigcap_{n=1}^{\infty} (xs)^n S = \emptyset$ and so $\bigcap_{n=1}^{\infty} (sx)^n S = \emptyset$ in similar fashion. Therefore sx and $xs \in T$, which implies that T is an ideal. Suppose now that S is commutative. Then T is completely prime, since otherwise there exist a and $b \notin T$ but $ab \in T$. We can choose $y \in \bigcap_{n=1}^{\infty} a^n S$ and $z \in \bigcap_{n=1}^{\infty} b^n S$. Then $yz \in a^n S \cdot b^n S \subseteq (ab)^n S$ for every natural number n, which implies $\bigcap_{n=1}^{\infty} (ab)^n S \neq \emptyset$. This is a contradiction.

COROLLARY 2.26. *Let S be a positively t.o. commutative semigroup. Then* $T = \{x: \bigcap_{n=1}^{\infty} x^n S = \emptyset\}$ *is a completely prime ideal, if nonempty.*

PROPOSITION 2.27. *Let S be a positively t.o. semigroup. If x is not an o-Archimedean element, then* $S_x = \{z: x^n < z$ *for every natural number* $n\}$ *is a nonempty convex completely prime ideal. If x and y are not o-Archimedean elements, then* $S_y < S_x$ *implies* $y > x$; $y > x$ *implies* $S_y \subset S_x$ *and* $S_{xy} = S_x \cap S_y$.

PROOF. S_x is an ideal, since, if $y \in S_x$ and $z \in S$, then $x^n < y \leq yz$ for every natural number n and so $yz \in S_x$. Similarly $zy \in S_x$. Suppose $ab \in S_x$ with $a \notin S_x$ and $b \notin S_x$. Then $x^n \geq a$ and $x^m \geq b$ for some natural numbers n and m and so $x^{n+m} \geq ab$, which contradicts that $ab \in S_x$. This establishes that S_x is completely prime. It can easily be verified that S_x is convex. Let $S_y < S_x$. If possible let $y \leq x$. Pick $t \in S_x \backslash S_y$. Then for some n, $y^n \geq t$. But $x^n < t$. Therefore $t \leq y^n \leq x^n < t$, which is absurd. Thus $y > x$.

Let $y > x$. If $z \in S_y$, then $y^n < z$ for every natural number n. Hence $x^n \leq y^n < z$, which implies $z \in S_x$. Thus $S_y \subset S_x$. To prove the last statement, let $t \in S_{xy}$. Then for every natural number n, $x^n \leq (xy)^n < t$ and also $y^n < t$. Hence $t \in S_x \cap S_y$. If $t \in S_x \cap S_y$, then $x^n < t$ and $y^n < t$ for every natural number n. We may assume that $x \leq y$. Then $(xy)^n \leq y^{2n} < t$, so that $t \in S_{xy}$. Thus $S_{xy} = S_x \cap S_y$.

The following two results, which are of independent interest, describe the nature of o-Archimedean and non-o-Archimedean elements.

PROPOSITION 2.28. *If e is a nonzero idempotent in a positively t.o. semigroup S, and $1 \neq x < e$, then x is non-o-Archimedean.*

PROOF. Let x be o-Archimedean. Since $x^n \leq e$ for every natural number n, $x^n = e$. Then by 2.1, e is a zero of S, which is a contradiction.

PROPOSITION 2.29. *Let $x \neq 1$ be an o-Archimedean element in a positively t.o. semigroup S. Then $\bigcap_{n=1}^{\infty} x^n S = \emptyset$ (or 0) and $\bigcap_{n=1}^{\infty} Sx^n = \emptyset$ (or 0) (if S contains 0).*

PROOF. Suppose $y \in \bigcap_{n=1}^{\infty} x^n S$. Then by positive order, $y \geq x^n$ for every natural number n. Hence $y = x^n$ since x is o-Archimedean. Then $x^n \in x^n S$ implies that $x^n = x^{n+1}$ is an idempotent. Since $x^n = 1$

implies $x = 1$ by 2.1, $x^n < y$ is impossible for any $y \in S$. Hence x^n is a maximal element, which is 0 by 2.1.

If S contains proper completely prime ideals, then the union Q of all these proper completely prime ideals is a completely prime ideal. Now we shall show how these ideals Q and A (set of all o-Archimedean elements) play an important role in describing the structure of arbitrary positively t.o. semigroups.

PROPOSITION 2.30. *If S is a positively t.o. semigroup, then $S\backslash Q$ is an o-Archimedean positively t.o. semigroup, where Q is as defined above.*

PROOF. If S contains 1, then $S\backslash Q = \{1\}$ and hence the result is trivially true. So assume that S does not contain 1. Since Q is completely prime, $S\backslash Q$ is a subsemigroup. Let $x,y \in S\backslash Q$. As noted in the proof of 2.9, $T_x = \{a\colon a \in S \text{ and } a^n \geq x \text{ for some } n\}$ is a completely prime ideal containing x. So if $T_x \neq S$, then $x \in T_x \subset Q$, which is not true. Thus $T_x = S$. Similarly $T_y = S$. So $x^n \geq y$ and $y^m \geq x$ for some natural numbers n and m. Thus $S\backslash Q$ is o-Archimedean.

THEOREM 2.31. *Let S be a right cancellative, positively t.o. semigroup. Then S is one of the following:*

(i) S is ζ-indecomposable;

(ii) $S = Q$, the union of all proper completely prime ideals;

(iii) $S \neq Q$ and $S = Q \cup \{x\}$, where x is an idempotent and a minimal element or for every $s \in S$ there exist an $x \in S$ and a natural number n, such that $s^n \geq x > s$ (and hence no idempotents in S).

PROOF. Let T be the set of all elements s in S such that $s^n \geq x > s$ for some natural number n and for some x in S. If $a \in T$ and $s \in S$, then $a^n \geq x > a$ for some $x \in S$ and for some n and so

$$(as)^{n+1} \geq a^n \cdot as \geq xas \geq xs > as$$

the last strictly inequality being possible by right cancellative condition. Hence as ε T. Clearly $a \neq a^2$. If, for any $s \in S$, $sa = (sa)^2$, then $sa \leq sa^n \leq sa(sa)^n = (sa)^{n+1} = sa$ and so $sa = sa^2$. By right cancellation $s = sa$. Then $sa = sasa = s^2a$ implies $s = s^2$. Therefore $as = as^2$ and $a = as = asa = a^2$, a contradiction. Hence $sa \neq (sa)^2$ and $sa < (sa)^2 \leq (sa)^3$. Thus $sa \in T$. Thus T is an ideal. T is completely prime. For, let $ab \in T$. For definiteness assume $a \geq b$. Then there exists an $x \in S$ and a natural number n such that $(ab)^n \geq x > ab$ and so $a^{2n} \geq (ab)^n \geq x > ab \geq a$, which implies $a \in T$. Let S be not type (i). Then S has proper completely prime ideals. As above assume that Q is the union of all proper completely prime ideals. Suppose that S is not type (ii); i.e., $S \neq Q$. Then $S = T$ or $S \neq T$. Now consider the case when $S \neq T$. Then $|S \backslash Q| = 1$ since if $x, y \in S \backslash Q$ and $x < y$, then by 2.29 there exists a natural number n such that $x^n \geq y > x$, so that $x \in T \subset Q$, which is a contradiction. Thus $S \backslash Q = \{x\}$ for some x in S. Since Q is completely prime, $S \backslash Q$ is a subsemigroup and so $x^2 = x \in S \backslash Q$. Now for every $y \in S$, $yx = yx^2$. So by right cancellative condition, $y = yx \geq x$. Thus x is a minimal element.

LEMMA 2.32. *In a positively t.o. semigroup S, every o-Archimedean element is greater than every non-o-Archimedean element. The set A of all o-Archimedean elements is a convex completely prime ideal, if nonempty and* $A = \bigcap S_x$, *where x ranges over all non-o-Archimedean elements and* $S_x = \{y : y \in S, x^n < y$ *for every natural number* $n\}$.

PROOF. Let a be o-Archimedean and b be non-o-Archimedean. Suppose $a < b$. Since $b^n < t$ for some $t \in S$ and for every natural number n, $a^n \leq b^n < t$ for every n, which is impossible. Thus $a > b$. Clearly convexity of A is evident. From 2.24, we note that A is a completely prime ideal. If $a \in A$ and if x is non-o-Archimedean, then from the above $x^n < a$ for every natural number n. Hence $a \in S_x$ and thus $A \subseteq \bigcap S_x$. Conversely if $y \in \bigcap S_x$ and $y \notin A$, then $y \in S_y$ and so $y^n < y$ for every n, which is impossible.

PROPOSITION 2.33. *Let S be a positively t.o. semigroup, with Q and A as defined above. Assume that S does not contain identity. Then the following are true:*

(i) If $e = e^2 \in Q$ and if $f = f^2 \in S \backslash Q$, then $e > f$.

(ii) If $e = e^2 \in Q$ and if $f = f^2 \in S \backslash Q$, then $e > x$ for every $x \in S \backslash Q$.

(iii) If x is a periodic element in Q, then $x > y$ for every $y \in S \backslash Q$.

(iv) $Q \backslash A$ is a subsemigroup.

(v) If S is commutative and x is non-o-Archimedean, then $S_x \subseteq Q$, where $S_x = \{y: x^n < y$ for every natural number $n\}$.

PROOF. (i) If $e < f$, then $f = ef$ by (vi) of 2.1, so that $f \in Q$, which is not true.

(ii) Since $S \backslash Q$ is o-Archimedean by 2.30, f is a maximal element by (ii) of 2.2. Since $e > f$ by (i), $e > b$ for every $b \in S \backslash Q$.

(iii) Since x is periodic, $x^m = x^n$ for some natural numbers m and n. Then by positive order we may assume that x^m is an idempotent. Suppose that $y \in S \backslash Q$ and $x < y$. Then $x^m \leq y^m$. Since $S \backslash Q$ is a subsemigroup by 2.30, $y^m \in S \backslash Q$. Hence the inequality $x^m \leq y^m$ is not possible by (ii) since $x^m \in Q$.

(iv) Let $x, y \in Q \backslash A$. If $xy \in A$, then x or $y \in A$ since A is completely prime by 2.24. Clearly $xy \in Q$ since Q is an ideal. Thus $xy \in Q \backslash A$.

(v) Let $y \in S_x$. If possible let $y \in S \backslash Q$ and $x^n \notin yS^1$ for every natural number n. Let H be the collection of all ideals which do not contain any power of x but contain yS^1. Then, by Zorn's lemma, there exists a maximal element P in H with respect to the ordering of set-inclusion. P is a completely prime ideal since if $ab \in P$ with a and $b \notin P$, then for some natural numbers m and n, we have $x^m \in P \cup aS^1$

and $x^n \in P \cup bS^1$ by the maximality of P. Hence $x^{m+n} \in (P \cup aS^1)(P \cup bS^1) \subseteq P$, which is not true. Since every proper completely prime ideal is contained in Q and $yS^1 \not\subseteq Q$ by supposition, P = S. But S contains all powers of x. Hence this contradiction proves $y \in Q$.

Since the only invertible element in a positively t.o. semigroup is the identity only, without loss of generality we may assume that these positively t.o. semigroups have no identity. We have proved in 2.10 that every positively t.o. semigroup is a semilattice of o-Archimedean semigroups. Since $S = A \cup Q\backslash A \cup S\backslash Q$, with the notation defined above and since A and $S\backslash Q$ are o-Archimedean subsemigroups, S is a disjoint union of a finite number of o-Archimedean semigroups, when once we prove that $Q\backslash A$ is an o-Archimedean subsemigroup. We do not know at present when $Q\backslash A$ is o-Archimedean though it is a subsemigroup by 2.33. We have examples for supporting this possibility. Let S be a semigroup generated by x_1, x_2, x_3 subject to the relations $x_i x_j = x_j x_i$ for $i \neq j$ and $x_i x_j = x_j$ if $i < j$. Order S by: $x_1 < x_1^2 < \ldots < x_2 < x_2^2 < \ldots < x_3 < x_3^2 < \ldots$. Here $Q\backslash A$ is the set of all powers of x_2, which is clearly an o-Archimedean subsemigroup. Also A, $S\backslash Q$, $Q\backslash A$ are all convex subsemigroups. In the next chapter we will show that all these subsemigroups are convex under some special hypothesis.

Let H be the family of all proper completely prime ideals of an arbitrary semigroup S. Then we can define a congruence ρ on S: $a\rho b$ if for every $A \in H$, either $a,b \in A$ or $a,b \notin A$. The congruence classes under ρ are called N-classes. These N-classes are ζ-indecomposable (or N-simple). S is a semilattice of these N-classes. If S is a positively t.o. semigroup, then each of these N-classes is o-Archimedean by 2.9. It may happen that these N-classes are always convex subsemigroups. For example let S be semigroup generated by x and y subject to the relation $xy = yx = y$ and ordered by: $x < x^2 < \ldots < y < y^2 < \ldots$. The infinite cyclic subsemigroups generated by x and y separately are the N-classes, which are convex. We do not know whether these N-classes are always convex subsemigroups. But if this happens, then every positively t.o. semigroup is a semilattice of convex o-Archimedean positively t.o. semigroups, which improves the representation theorem of Saito given in [3]. Now we provide a sufficient condition for a N-class to be a convex subsemigroup.

PROPOSITION 2.34. *Let N be a N-class of a positively t.o. semigroup. If N is a nilsemigroup (or equivalently N contains a maximal element), then N is a convex subsemigroup.*

PROOF. Let $a < b < c$ with $a,c \in N$. If e is the zero of N, then $a^n = e = c^n$ for some natural number n. Now $e = a^n \leq b^n \leq c^n = e$. Hence $b^n = e$ and therefore e belongs to the N-class containing b. Since N-classes are disjoint, $b \in N$. Thus N is a convex subsemigroup.

In the following theorems we shall show that some classes of cancellative, commutative, positively t.o. semigroups can be built from groups and additive semigroups of non-negative integers.

An element a in a semigroup S is called *central* if it commutes with every element of S.

LEMMA 2.35. *Let a be a central cancellable element in a semigroup S. If $a = ab$ for some $b \in S$, then b is the identity of S.*

PROOF. Since $a = ab$ implies $ab = ab^2$, $b = b^2$. Now if $y \in S$, $ya = yab = (yb)a$, which implies $y = yb$. But $(ab)y = ay$, so that $(by)a = ya$ and $by = y$. Thus b is the identity of S.

LEMMA 2.36. *Let S be a positively t.o. semigroup in which Q* (the intersection of all completely prime ideals) is nonempty. Then S contains o-Archimedean elements.*

PROOF. Suppose that S does not contain o-Archimedean elements. Let $y \in Q^*$. Since y is non-o-Archimedean, S_y is a nonempty completely prime ideal by 2.27. Hence $y \in S_y$; i.e., $y^n < y$ for every natural number n, which is impossible.

The following lemma is crucial for the development of our theorems. The proof of this lemma is essentially that of Tamura's theorem [1;53].

LEMMA 2.37. *Let S be a semigroup containing a central cancellable element a such that* $\bigcap_{n=1}^{\infty} a^n S = \emptyset$, $a \notin aS$, *and for every x in S,* $a^m \in xS$ *for some natural number n. If N is the additive semigroup of nonnegative integers; G is a group and* $I\colon G \times G \to N$ *is a function satisfying:*

(i) $I(\alpha,\beta) + I(\alpha\beta,\gamma) = I(\alpha,\beta\gamma) + I(\beta,\gamma)$ *where* α, β, *and* γ *are the elements of G;*

(ii) $I(\varepsilon,\varepsilon) = 1$, *where* ε *is the identity of G (conditions (i) and (ii) imply* $I(\alpha,\varepsilon) = I(\varepsilon,\alpha) = 1$ *for every* $\alpha \in G$), *then S is isomorphic with the semigroup* $N \times G$, *where the multiplication in* $N \times G$ *being defined by:*
$(m,\alpha)(n,\beta) = (m + n + I(\alpha,\beta),\alpha\beta)$.

PROOF. Let $T_n = a^n S \backslash a^{n+1} S$, $n = 1,2,\ldots$ with $T_0 = S \backslash aS$. If $x \notin aS$, write $x = a^0 x$. If $x \in aS$, since $\bigcap a^n S = \emptyset$, we must have $x \in T_n \backslash T_{n+1}$ for some n. Hence $x = a^n z$ where $z \notin aS$. Thus every $x \in S$ can be written as $x = a^n z$, $z \notin aS$, where $n = 0,1,2,\ldots$.

This representation is unique. For, let $a^n z = a^m w$ with $z,w \notin aS$. Since a is cancellable $z \in aS$ or $w \in aS$ according as $n < m$ or $n > m$ respectively. Since this is impossible, $n = m$. Then again by cancellative condition of a, $z = w$.

Define a relation σ on S by: $x\sigma y$ iff $a^m x = a^n y$ for some natural numbers $m,n \geq 0$. Since a is central, σ is a congruence. Denote the classes of σ by S_α, $\alpha \in G$. G is a semigroup by defining the product

$\alpha\beta$ in G is the index of the class containing xy where $x \in S_\alpha$ and $y \in S_\beta$. All positive powers of a constitute a single σ-class. For, if $x\sigma a$ and if x is not a power of a, then $x = a^m z$, $z \notin aS$. Now $a^p(a^m z) = a^q(a)$ for some natural numbers $p,q \geq 0$. If $p + m \leq q$, then $z = a^{q-p-m+1}$ since a is cancellable and so x is a power of a, which is not true. If $p + m > q$, then $a \in aS$, which is again not true by hypothesis. Denote the class containing all positive powers of a by S_ε. Then ε is the identity of G. We observe that G is a group. For, let $\alpha \in G$. If $x \in S_\alpha$, then, by hypothesis, $a^m = xy$ for some natural number m and for some $y \in S$, so that $a\sigma xy$ and thus if S_β is the σ-class containing y, then $S_\varepsilon = S_\alpha S_\beta$ and $\varepsilon = \alpha\beta$. By the same reason $\varepsilon = \beta\gamma$ for some $\gamma \in S$. Thus $\varepsilon = \alpha\beta = \beta\alpha$ and hence G is a group.

Since every $x \in S$ can be written uniquely in the form $x = a^n z$ for some $z \in T_0 = S\backslash aS$, it follows that every x is σ-related to exactly one element of T_0 and thus T_0 intersects each σ-class S_α in a single element, say u_α. Then $S_\alpha = \{a^n u_\alpha : n > 0, \alpha \in G\}$.

By definition $u_\alpha u_\beta \in S_{\alpha\beta}$ and so $u_\alpha u_\beta \sigma u_{\alpha\beta}$. Then we have $a^n u_\alpha u_\beta = a^m u_{\alpha\beta}$ for some natural numbers n and m. Since a is cancellable, if $n > m$, then $u_{\alpha\beta} \in aS$, which is not true. Thus $u_\alpha u_\beta = a^n u_{\alpha\beta}$ for some n and because of unique representation, this n is unique. Define a map $I: G \times G \to N$ by $I(\alpha,\beta) = n$ where $u_\alpha u_\beta = a^n u_{\alpha\beta}$.

Since $(u_\alpha u_\beta)u_\gamma = u_\alpha(u_\beta u_\gamma)$, for any $\alpha,\beta,\gamma \in G$, we have

$$a^{I(\alpha,\beta)} u_{\alpha\beta} u_\gamma = u_\alpha a^{I(\beta,\gamma)} u_{\beta\gamma} = a^{I(\beta,\gamma)} u_\alpha u_{\beta\gamma}$$

and so $a^{I(\alpha,\beta)} a^{I(\alpha\beta,\gamma)} u_{\alpha\beta\gamma} = a^{I(\beta,\gamma)} a^{I(\alpha,\beta\gamma)} u_{\alpha\beta\gamma}$. Then by uniqueness of the representation we have

$$I(\alpha,\beta) + I(\alpha\beta,\gamma) = I(\beta,\gamma) + I(\alpha,\beta\gamma)$$

Since $u_\varepsilon u_\varepsilon = a^{I(\varepsilon,\varepsilon)} u_{\varepsilon^2} = a^{I(\varepsilon,\varepsilon)} u_\varepsilon$ and $a = u_\varepsilon$, we have $a^2 = a^{I(\varepsilon,\varepsilon)+1}$, which implies $I(\varepsilon,\varepsilon) = 1$, because a is cancellable and $a \notin aS$.

Since $a = u_\varepsilon$ is central, for every $\alpha \in G$ we must have $a^{I(\alpha,\varepsilon)}u_{\alpha\varepsilon} = u_\alpha u_\varepsilon = u_\varepsilon u_\alpha = a^{I(\varepsilon,\alpha)}u_{\varepsilon\alpha}$ and so $a^{I(\alpha,\varepsilon)}u_\alpha = a^{I(\varepsilon,\alpha)}u_\alpha$. Then, by uniqueness of the representation of elements, $I(\alpha,\varepsilon) = I(\varepsilon,\alpha)$.

Define a mapping ψ by:

$$x\psi = (n,\alpha) \text{ where } x = a^n u_\alpha \in S$$

Because of the uniqueness of representation of elements, this ψ establishes semigroup isomorphism between S and $N \times G$.

LEMMA 2.38. *Let S be a semigroup in which every right ideal contains an ideal. If $a \in P^*$ (the intersection of all prime ideals) then for every x in S there exists a positive integer m such that $a^m \in xS$.*

PROOF. Suppose that no power of a is in xS. Since xS contains an ideal, say T, no power of a is in T. Then the collection F of all ideals containing T and not containing any power of a is nonempty. So by Zorn's lemma, there exists an ideal P, which is a maximal element in F with respect to the ordering of set-inclusion. Now we claim that P is a prime ideal, which establishes the conclusion. Suppose there exist ideals B and C such that $BC \subset P$, without B and C being contained in P. By maximality of P, there exist natural numbers m and n such that $a^m \in P \cup B$ and $a^n \in P \cup C$, so that $a^{m+n} \in P$, which is not true.

LEMMA 2.39. *If right ideals in a semigroup S are two-sided, then prime ideals of S are completely prime ideals.*

PROOF. Let P be a prime ideal in S and $xy \in P$. Since $S^1xS^1S^1yS^1 \subset S^1xyS^1 \subset P$, we must have then $S^1xS^1 \subset P$ or $S^1yS^1 \subset P$, which implies that x or y is in P. Hence P is completely prime.

THEOREM 2.40. *Let S be a ζ-indecomposable, positively t.o. semigroup containing a central cancellable element. If $|S| > 1$ and if every right ideal in S is two-sided, then S is isomorphic with $N \times G$ as described in 2.37.*

PROOF. By 2.9, S is o-Archimedean. Then the only possible idempotents are 0 and 1 by (ii) of 2.2. Let a be the central cancellable element in S. If S contains 0, then S is a nilsemigroup by 2.5. Hence $a^n = 0$ for some natural number n, which implies $a = 0$ and so $|S| = 1$. Since the only invertible element in S is 1 itself by (iii) of 2.1, $S\backslash 1$ is a proper completely prime ideal, which contradicts the hypothesis. So S does not contain 1. Then by 2.29, $\bigcap_{n=1}^{\infty} a^n S = \emptyset$ and by 2.35, $a \notin aS$. Since S is ζ-indecomposable, S has no proper completely prime ideals and hence by 2.39, S has no proper prime ideals. Therefore $a \in P^* = S$, which implies by 2.38, that for every $x \in S$, $a^m \in xS$. Then by 2.37, the result follows.

COROLLARY 2.41. *Let S be a commutative ζ-indecomposable positively t.o. semigroup. If $|S| > 1$ and if S contains cancellable elements, then S is isomorphic with $N \times G$ as described in 2.37, where G is an abelian group.*

At present we are not able to prove the converse of 2.40 or 2.41. One may doubt whether there exist groups G and I-functions such that $N \times G$ satisfies the above hypotheses. We furnish now some examples.

EXAMPLE 1. Let G be an one-element group $\{\varepsilon\}$. Define order by: $(m,\varepsilon) > (n,\varepsilon)$ if $m > n$.

EXAMPLE 2. Let $G = \{\varepsilon, x: x^2 = \varepsilon\}$. Define order by:

.....

.....

$\ldots\ldots(m,x) > (m - 1,\varepsilon) > (m - 1,x) > (m - 2,\varepsilon) > (m - 2,x) >$
$\ldots > (1,x) > (0,\varepsilon) > (0,x).$

THEOREM 2.42. *Let S be a positively t.o. semigroup not containing identity. Then S is isomorphic with N × G as described in 2.37 if S satisfies the conditions:*

(i) $|S| > 1$;

(ii) every right ideal in S is two-sided;

(iii) P contains a central cancellable o-Archimedean element.*

PROOF. Let a be a central cancellable o-Archimedean element in P*. By 2.35, $a \notin aS$. By 2.9, $\bigcap a^n S = \emptyset$. For every $x \in S$, $a^m \in xS$ for some natural number m by virtue of 2.38. Hence by 2.37, the conclusion is evident.

COROLLARY 2.43. *Let S be a commutative cancellative positively t.o. semigroup without identity. If* $|S| > 1$ *and if* $P^* \neq \emptyset$, *then S is isomorphic with N × G as described in 2.37.*

PROOF. Since $P^* \neq \emptyset$, S contains o-Archimedean elements by 2.36. But the set A of all o-Archimedean elements is a completely prime ideal by 2.32. Thus P* contains central cancellable o-Archimedean elements. Hence by 2.42, this corollary is evident.

The example which satisfies the hypothesis of the Corollary 2.43 is N × G where G is the additive group of integers and the I-function defined by:

$$I(n,m) = 0 \text{ if } n,m > 0;\ I(n,-m) = 0 \text{ if } 0 < n \leq m \text{ and}$$

in all other cases the I-function takes the value 1.

To prove the converse, one has to know whether $N \times G$ admits a positive order always and under what conditions that $N \times G$ satisfies the hypothesis of Theorem 2.42. This is an open problem.

CHAPTER 3

NATURALLY TOTALLY ORDERED SEMIGROUPS

A semigroup S is said to be *right naturally totally ordered* (n.t.o.) if S is a positively t.o. semigroup and for $a,b \in S$, $a < b$ iff $b = as$ for some $s \in S$. Dually left n.t.o. semigroups can be defined. S is a n.t.o. semigroup if S is a right as well as left n.t.o. semigroup. There exist right n.t.o. semigroups which are not left n.t.o. and hence not n.t.o. Let S be a semigroup generated by two symbols x and y subject to the relation $y = xy$. Then every element of S is of the form $y^m x^n$, where m and n are nonnegative integers. S is ordered by:

$$x < x^2 < \ldots < x^n < \ldots < y < yx < yx^2 \ldots < y^2 < y^2x < \ldots$$

It can easily be verified that S is a right n.t.o. semigroup but it is not a left n.t.o. semigroup since $y \notin Sx$ and $x \notin Sy$.

The first classical problem considered in literature is to find conditions when a t.o. semigroup is a subsemigroup of the additive group of real numbers under usual order. Hölder and Clifford utilized the n.t.o. condition to obtain this characterization. N.t.o. semigroups also naturally occur as positive cones of t.o. groups and every cancellative positively t.o. semigroup with quotient condition, in particular, commutative cancellative positively t.o. semigroups, can be embedded in n.t.o. semigroups. In this chapter, along with the elucidation of these results, we shall discuss the structure and properties of n.t.o. semigroups. The structure is completely determined when the n.t.o. semigroup is o-Archimedean or finitely generated or commutative, cancellative semigroup containing o-Archimedean and non-o-Archimedean elements.

PROPOSITION 3.1. *For a right n.t.o. semigroup S, the following are true:*

(i) The principal right ideal $xS^1 = \{y \in S: y \geq x\}$;

(ii) Right ideals are convex;

(iii) Right ideals are linearly ordered under set inclusion;

(iv) Right ideals are two-sided;

(v) S contains a minimal element iff $S = x \cup xS$. *If S is finitely generated as a right ideal or if S is finitely generated, then S contains a minimal element;*

(vi) If $S \neq S^2$, *then* $S \backslash S^2 = \{x\}$ *and x is a minimal element;*

(vii) If S is o-Archimedean and if S is not right or left cancellative, then S contains a maximal element provided S does not contain identity;

(viii) Prime ideals are completely prime;

(ix) If S is not o-Archimedean, then the set A of all o-Archimedean elements is a completely prime ideal contained in every completely prime ideal.

<u>*PROOF*</u>. (i) By positive order $y \in xS^1$ implies $y \geq x$. If $y > x$, then $y \in xS$ by right n.t.o. order. Thus xS^1 is of the above form.

(ii) and (iii) follow readily from definition.

(iv) If A is a right ideal, then for any $a \in A$ and $s \in S$, $a \leq sa$ by positive order, so that $sa \in aS^1 \subseteq A$.

(v) If $S = x \cup xS$, then clearly x is a minimal element by positive order. Conversely if x is a minimal element and if $y (\neq x) \in S$, then $x < y$ implies $y \in xS$ and hence $S = x \cup xS$. Clearly the minimal among the finite set of generators is the minimal element in S because of positive order.

(vi) Observe that if $x,y \in S\backslash S^2$ and if $x \neq y$, then $x \in yS \subset S^2$ or $y \in xS \subset S^2$ by (iii). Thus $S\backslash S^2$ contains a unique element, say x. Since $x \in yS$ implies $x \in S^2$, we must have by (iii) $y \in xS^1$ for any $y \in S$ and thus $y \geq x$ by positive order. Hence x is a minimal element.

(vii) We recall that the conditions $a = ab$ or $a = ba$ for some a and b in S imply that a is a maximal element in o-Archimedean case by (i) and (ii) of 2.2. Now assume that S is not right cancellative; i.e., there exists a,b,c in S such that $ba = ca$ with $b < c$. By right n.t.o. condition, $c = bx$ for some $x \in S$. If $a = xa$, then from the above a is the maximal element. If $a < xa$, then $xa = at$ for some $t \in S$. Therefore $u = ca = bxa = bat = cat = ut$. This implies that u is a maximal element. Similar proof holds in left cancellative case.

(viii) Combine (iv) and 2.39.

(ix) By 2.32 A is a completely prime ideal. Let P be a completely prime ideal and $y \in P$. If $x \in A$, there exists a natural number n such that $x^n \geq y$. If $y \neq x^n$, then $x^n \in yS \subset P$. So in either case $x^n \in P$ and so $x \in P$. Thus $A \subset P$.

THEOREM 3.2. *Let S be an o-Archimedean right n.t.o. semigroup. If S is not left cancellative or right cancellative, then S is a nilsemigroup. If $S \neq S^2$ or S is finitely generated as a right ideal or S is finitely generated or S contains a minimal element, then S is a cyclic semigroup.*

PROOF. In the first case by (vii) of 3.1, S contains a maximal element and so S is a nilsemigroup by 2.5.

By (v) and (vi) of 3.1, all the conditions in the second case imply that S contains a minimal element, say x. Then by (v) of 3.1, $S = x \cup xS$. Now if y is not a power of x, then $y = x^n s_n$ for every natural number n, which implies $y > x^n$ by positive order. This is absurd. Thus every element of S is a power of x.

We prove now that the existence of minimal elements is also related to the presence of maximal ideals in S.

An element e in a semigroup S is called a *left identity* if $ex = x$ for every x in S.

PROPOSITION 3.3. *Let S be a right n.t.o. semigroup. Then the following are equivalent:*

(i) S contains a minimal element x;

(ii) $S = x \cup xS$ for some $x \in S$;

(iii) S contains a unique maximal right ideal.

Furthermore if $S = S^2$, then the above three equivalent conditions are also equivalent to the existence of left identity.

PROOF. (i) $\Rightarrow$ (ii) is true by (v) of 3.1.

(ii) $\Rightarrow$ (iii). We claim that $S\backslash x$ is a maximal right ideal, which is unique by (iii) of 3.1. It suffices to prove that $S\backslash x$ is a right ideal. Suppose that there exists an $y \in S\backslash x$ and $s \in S$ such that $ys \notin S\backslash x$. Then $ys = x$ and so $x \geq y$ by positive order. By the same reason, since $S = x \cup xS$, $y \geq x$. Hence $y = x$, which is a contradiction.

(iii) $\Rightarrow$ (i). If M is a maximal right ideal of S, then there exists an element $x \in S\backslash M$. By right n.t.o. condition, we must have then M is properly contained in $x \cup xS$. Hence $S = x \cup xS$, which implies by positive order that x is a minimal element.

If $S = S^2$ and if x is a minimal element, then $x = yz$ for some y,z in S, which implies $x \geq y$ and $x \geq z$ by positive order. Hence $x = y = z$ and so $x = x^2$. Then for any $y \neq x$ in S, $x < y$ and so $y = xy$. Thus x is a left identity. Conversely if e is the left identity of S, then for every y in S, we have $y = ey$ and so $y \geq e$. Thus e is a minimal element.

We notice in the above proof that the unique maximal right ideal is $S\backslash x$ if x is the minimal element. So $S\backslash x$ is also the unique maximal left ideal and maximal two-sided ideal. Since right Noetherian semigroups contain maximal right ideals, we have

COROLLARY 3.4. *Right Noetherian right n.t.o. semigroups contain minimal elements.*

By virtue of 3.2 and 3.4, the following is evident:

COROLLARY 3.5. *If an o-Archimedean right n.t.o. semigroup without identity is right Noetherian, then it is a cyclic semigroup.*

The set of all positive real numbers under addition and with the usual ordering is a non-Noetherian n.t.o. semigroup. But the infinite cyclic semigroup $\{x, x^2, \ldots\}$ with $x^n < x^{n+1}$ for every natural number n is a (right) n.t.o. semigroup which is Noetherian. In fact the latter is the exact characterization of some classes of Noetherian semigroups.

THEOREM 3.6. *Let S be a right cancellative right n.t.o. semigroup. Then the following are equivalent:*

(i) S is an infinite cyclic semigroup;

(ii) $S \neq S^2$ and S is a left Noetherian semigroup;

(iii) $S \neq S^2$ and S is a right Noetherian semigroup;

(iv) $S \neq S^2$ and between any two elements, there are at most a finite number of elements;

(v) $\bigcap_{m=1}^{\infty} S^n = \emptyset$.

PROOF. Trivially (i) implies all the other conditions.

(ii) → (i). Since $S \neq S^2$, $S = x \cup xS$ for some $x \in S \setminus S^2$ by (v) and (vi) of 3.1. By right cancellative condition no two powers of x will be the same. Suppose that some y in S is not a power of x. Then $x < y$ and so $y = xs_1$ where $s_1 \neq x$. Now $x < s_1$ implies $s_1 = xs_2$ for some s_2 in S. $s_1 \neq s_2$ since otherwise $s_1 = xs_1 = x^2 s_2$, which implies $x = x^2$ by right cancellative condition. Then $S = S^2$, which is not true. Thus we have a sequence $\{s_n\}$ such that $s_n = xs_{n+1}$ and no two s_n's are the same and also a chain of left ideals, namely, $S^1 s_1 \subset S^1 s_2 \subset \ldots$. Then by left Noetherian condition, there exists a natural number n such that $S^1 s_n = S^1 s_{n+1}$. Hence $s_{n+1} = ts_n$ for some $t \in S$ and $s_{n+1} = txs_{n+1} = (tx)^2 s_{n+1}$, which implies $tx = (tx)^2$ by right cancellative condition. Then for every $z \in S$, we have $z(tx) = z(tx)^2$ and so $z = z(tx) \in S^2$. Thus $S = S^2$, which is a contradiction. Thus every element of S is a power of x.

(iii) $\Rightarrow$ (i). As in the proof of (ii), $S = x \cup xS$ for some x in $S \backslash S^2$. Suppose that there exists an element y which is not a power of x. Then $y = x^i s_i$ where $s_i = xs_{i+1}$. Since S is positively ordered, $s_{i+1} \leq xs_{i+1} = s_i$ and so, by right n.t.o. condition, $s_i S^1 \subset s_{i+1} S^1$. Then by right Noetherian condition $s_i S^1 = s_{i+1} S^1$ for some i. This implies $s_i = s_{i+1}$ by positive order condition. Hence $s_i = xs_{i+1} = xs_i$ and $x = x^2$ by right cancellation. This is absurd since $x \notin S^2$. Thus every element of S is a power of x and no two powers of x coincide, since otherwise, $x^i = x^{i+1}$ implies $x = x^2 \in S^2$.

(iv) $\Rightarrow$ (i). As indicated in the above proof, if $y \neq x^n$ for any n, then $s_1 > s_2 > \ldots > x$, which is not possible by hypothesis. Hence the conclusion is evident.

(v) $\Rightarrow$ (i). Since $\bigcap_{n=1}^{\infty} S^n = \emptyset$, $S \neq S^2$. As before $S = x \cup xS$ where $S \backslash S^2 = \{x\}$. Let $y \in S$. Since $\bigcap_{n=1}^{\infty} S^n = \emptyset$, there exists a natural number r such that $y \in S^r \backslash S^{r+1}$. But $S^r = x^{r-1}S$. Therefore $y = x^{r-1}s$ where $s = x$ or $s = xt$. If $s = xt$, $y \in x^r S = S^{r+1}$. Hence $y = x^r$.

In contrast to the above theorem, the semigroup $\{1, x, x^2, \ldots\}$ with $1 < x^n < x^{n+1}$ for every natural number n is a globally idempotent Noetherian cancellative (right) n.t.o. semigroup. But every left Noetherian cancellative, right n.t.o. globally idempotent semigroup need not be of this form. For example, let S be a semigroup generated by x and y. Let $y = xy$ and $x = x^2$. Define the order by:

$$x < y < yx < y^2 < y^2 x < \ldots$$

If the Noetherian condition is dropped in the above theorem, one can show a weaker conclusion as follows:

THEOREM 3.7. *Let S be a nonglobally idempotent right n.t.o. semigroup. If S is left cancellative or does not contain idempotents or* $S\backslash S^2$ *contains a right cancellative element, then either S is o-isomorphic with an infinite cyclic semigroup* $\langle x \rangle$ *with* $x^n < x^{n+1}$ *for every natural number n or there exists an o-homomorphism of S onto the above infinite cyclic semigroup* $\langle x \rangle$ *adjoined with zero.*

PROOF. Since $S \neq S^2$, $S = xS^1$ for some x in S by (v) and (vi) of 3.1. If S satisfies either of the first two conditions, then the conclusion follows from 2.22. Let now $S\backslash S^2$ contain a right cancellable element, say a. Clearly $Z = \{b \in S: xa = xb$ for some x,y in S with $x \neq y\}$ is a right ideal and hence a two-sided ideal by (iv) of 3.1. If $Z = \emptyset$, S is right cancellative and hence the conclusion is evident by 2.22. Let $Z \neq \emptyset$. Z is clearly a convex ideal by (ii) of 3.1. $\bar{S} = S/Z$ can be totally ordered in the usual way as before. $\bar{S} \neq \bar{S}^2$ since, $\bar{a} \neq \bar{0}$ and $\bar{a} = \bar{y}\bar{t}$ implies $a = yt \in S^2$, which is a contradiction. Hence $\bar{a} \in \bar{S}\backslash\bar{S}^2$. Since S is a right n.t.o. semigroup, $\bar{S}$ is also a right n.t.o. semigroup. The proof is now completed by showing that $\bar{S}$ is a right cancellative semigroup adjoined with zero since $\bar{S}$ then does have the above mapping property and the composition map of the natural homomorphism of S onto $\bar{S}$ and the map of $\bar{S}$ onto an infinite cyclic semigroup with zero provides the necessary conclusion. Suppose $\bar{x}\bar{y} = \bar{0}$ in $\bar{S}$ with $x,y \notin Z$. Then $xy \in Z$ and so there exist b and c in S with $b \neq c$ such that $bxy = cxy$. Since $x,y \notin Z$, this implies $b = c$, which is a contradiction. Therefore $\bar{x}\bar{y} \neq \bar{0}$. Let $\bar{x}\bar{y} = \bar{t}\bar{y} \neq \bar{0}$ and $x,y,t \notin Z$. Then $xy = ty$ or $xy \in Z$. Since $y \notin Z$, $x = t$ and hence $\bar{x} = \bar{t}$. The other condition $xy \in Z$ is impossible since $x,y \notin Z$, as before.

The property (ii) in 3.1, that every right ideal is convex, exactly characterizes right n.t.o. semigroups among positively ordered semigroups. For, if $x < y$, then $x < y \leq xy$ by positive order. Since xS^1 is convex, $y \in xS^1$ and so $y \in xS$.

For a certain class of monoids, the properties (iii) and (iv) in 3.1 provide a n.t.o. structure.

An element x in any semigroup S is called a *right (left) unit* if $yx = 1$ ($xy = 1$) for some y in S. x is called a *unit* if it is both a right and left unit.

THEOREM 3.8. *Let S be a monoid satisfying the conditions (iii) and (iv) in 3.1 and containing 1 as the only right unit. Then S can be endowed with a right n.t.o. structure if S is either left cancellative or S satisfies the property:* $\bigcap_{n=1}^{\infty} Sx^n = \emptyset$ *for every* $x \neq 1$.

<u>*PROOF*</u>. Set $a < b$ if $b \in aS$. Clearly this relation is reflexive and transitive. Suppose $a < b$ and $b < a$. Then $b = as$ and $a = bt$ for some $s,t \in S$. Hence $a = ast$ and also $a \in \bigcap_{n=1}^{\infty} S(st)^n$. This implies $st = 1$, so that $s = 1$ and $a = b$. Thus the antisymmetric condition of this order can be established. Because of the condition (iii) of 3.1, the linearity of this order is evident. For $a,b \in S$, we have $ab \in aS$ and so $a < ab$. By (iv) we have $ab \in Sb \subset bS$ and so $b < ab$. Thus S is positively ordered. To show the compatibility of order, assume $a < b$. Then $b \in aS$ and so for every $x \in S$ we have $bx \in aSx \subset axS$ since the right ideal xS is two-sided by the condition (iii). Hence $ax \leq bx$. Since $xb \in xaS$, $xa \leq xb$.

COROLLARY 3.9. *Let S be a semigroup without identity, in which the right ideals are two-sided and the right ideals form a chain under set-inclusion. Then S can be endowed with a right n.t.o. structure if S is either a left cancellative semigroup without idempotents or S satisfies the property* $\bigcap_{n=1}^{\infty} Sx^n = \emptyset$ *for every x in S.*

PROOF. If S^1 is a right n.t.o. monoid, then it is clear that S is also a right n.t.o. semigroup. So, by virtue of 3.8, it suffices to prove that S^1 is either left cancellative or $\bigcap_{n=1}^{\infty} S^1x^n = \emptyset$ for every $x \neq 1$. Now $a \cdot 1 = ac$ for $a,c \in S$ implies that c is an idempotent in S if S is left cancellative, which is not true. The implication of the second condition is evident.

THEOREM 3.10. *Let S be a left cancellative monoid in which the right ideals are two-sided and the only right or left unit is 1. If the left ideals form a chain under set-inclusion, then S can be endowed with a n.t.o. structure.*

PROOF. Set $a < b$ if $b \in Sa$. This implies that $b \in Sa \subset aS$, since right ideals are two-sided. Then the result follows as in the proof of Theorem 3.8.

In the following one can note the frequent occurrence of n.t.o. semigroups. In fact some classes of semigroups can always be embeddable in n.t.o. semigroups.

DEFINITION. The *positive cone* of a t.o. group with identity e is the set of all elements greater than e.

THEOREM 3.11. *The positive cone of a t.o. group is a n.t.o. semigroup.*

PROOF. Let S be the positive cone of a t.o. group with identity e. Let $a,b \in S$ and $a < b$. If a^{-1} is the inverse of a and $a^{-1}b < e$, then $b = a(a^{-1}b) \leq ae = a < b$, which is not true. So $a^{-1}b \in S$ and similarly $ba^{-1} \in S$. But $b = a(a^{-1}b) = (ba^{-1})a$ and thus S is a n.t.o. semigroup.

THEOREM 3.12. *Let S be a left and right cancellative positively t.o. semigroup satisfying the quotient condition and not containing identity. Then S can be embedded in a n.t.o. monoid.*

PROOF. By noting that right and left cancellative semigroups can have no idempotents other than identity, it follows easily that S^1 is a cancellative monoid. Since S^1 also satisfies the quotient condition, S^1 admits a group G of quotient of 1.2. Now define the order in S^1 by prescribing $s \geq 1$ for every $s \in S$ and by preserving the original order for the elements of S. Extend the order to G as described in Chapter 1. Let $P = \{x \in G: x \geq 1\}$. Then S is embedded in P. Clearly P is a monoid which can be shown a n.t.o. monoid as in 3.11.

COROLLARY 3.13. *Every cancellative commutative positively t.o. semigroup not containing identity can be embedded in a n.t.o. monoid.*

It is not always possible to embed positively t.o. semigroups in n.t.o. semigroups. For example consider the semigroup S generated by two elements a and f with $f^2 = f$ and ordered by: $f < a < af < a^2 < a^2f < \ldots < a^n < a^nf < \ldots$. If S is embedded in a n.t.o. semigroup, then $f < a$ should imply $a \in fS$ and hence $a = fa$, which is not true. The question when a positively t.o. semigroup is embeddable in a n.t.o. semigroup is an open problem.

So far we have studied the properties of right n.t.o. semigroups and how they occur in general. Now we shall describe the construction of finitely generated right n.t.o. semigroups.

THEOREM 3.14. *Let S be a finitely generated right n.t.o. semigroup. Assume that $\{x_1, x_2, \ldots, x_n\}$ is a minimal generating set with $x_1 < x_2 < \ldots < x_n$. Then $S = x_1S^1$ and S is one of the following forms:*

(i) $S = \{x_i^r : r = 1,2,\ldots,n\}$ *with* $x_i x_j = x_j = x_j x_i$ *for* $i < j$ *and ordered by:* $x_i^{r-1} \leq x_i^r$ *and* $x_i^r < x_{i+1}$ *for every natural number* r *and for* $i = 1,2,\ldots,n$.

(ii) Every element of S *is of the form* $x_n^{\alpha_n} x_{n-1}^{\alpha_{n-1}} \ldots x_1^{\alpha_1}$; $x_i x_j = x_j$ *for* $i < j$ *where* $i,j = 1,2,\ldots,n$ *and* S *is ordered by:*

$$x_1 \leq x_1^2 \leq \ldots \leq x_1^r \ldots < x_2 \leq x_2 x_1 \ldots \leq x_2 x_1^r \leq \ldots$$
$$\leq x_2^2 \leq x_2^2 x_1 \ldots \leq x_2^r x_1^r \leq \ldots \leq x_2^3 \leq x_2^3 x_1 \leq \ldots$$
$$\ldots$$
$$< x_3 \leq x_3 x_1 \leq \ldots \leq x_3 x_1^r \leq \ldots \leq x_3 x_2 \leq x_3 x_2 x_1 \leq \ldots$$
$$\leq x_3 x_2 x_1^r \leq \ldots \leq x_3 x_2^2 \leq x_3 x_2^2 x_1 \leq \ldots$$
$$\leq x_3^2 \leq x_3^2 x_2 \leq \ldots$$
$$\ldots$$
$$\ldots$$
$$< x_4 \leq x_4 x_1 \leq \ldots$$
$$\ldots$$
$$\ldots$$
$$< x_n \leq x_n x_1 \leq \ldots \leq x_n x_1^r \leq \ldots$$
$$\leq x_n x_2 \leq x_n x_2 x_1 \leq x_n x_2 x_1^2 \leq \ldots$$
$$\leq x_n x_2^2 \leq \ldots$$
$$\ldots .$$

Conversely if S *is one of the above two forms, then* S *is a right n.t.o. semigroup.*

PROOF. Let S be a right n.t.o. semigroup. Since $x_1 < x_i$ for $i = 2,\ldots,n$, $x_i \in x_1 S$ and hence $S = x_1 S^1$. If i and j are natural numbers such that $i < j$, then $x_i < x_j$ and so $x_j = x_i s$ for some s in S. In the expression of S no x_k with $k > j$ can occur, for otherwise, by positive order, $x_j \geq x_k$. This is impossible since $x_k > x_j$. If every x_k occurring in the expression of s satisfies the property $k < j$, then x_j is the product of other generators and so can be omitted. Hence there exists one x_k in the expression of s with $k = j$.

Thus $x_j \geq s \geq x_j$ and so $s = x_j$. Hence $x_j = x_i x_j$. This clearly implies that $x_i^n x_j^m = x_j$ for all natural numbers n and m. Then by positive order we have $x_i^n < x_j$ for $i < j$ for every natural number n and for $j = 2,3,\ldots,n$. Now it is evident that every element is of the form $x_n^{\alpha_n} x_{n-1}^{\alpha_{n-1}} \ldots x_1^{\alpha_1}$. Using the condition of positive order, it can be shown that S takes the first prescribed form if S is commutative and the second form if S is non-commutative. The converse is evident.

COROLLARY 3.15. *If S is a right cancellative, finitely generated right n.t.o. semigroup and if* $|S| > 1$, *then S is an infinite cyclic semigroup possibly adjoined with identity.*

PROOF. Suppose the minimal number n of generators be greater than and equal to 2. Then by Theorem 3.14, $x_n = x_1 x_n = \ldots = x_{n-1} x_n$, which implies $x_1 = x_2 = \ldots = x_{n-1}$ by right cancellative condition. Thus S can be generated only by two elements, say x_1 and x_2. Then $x_2 = x_1 x_2$ implies $x_2 = x_1 x_2 = x_1^2 x_2$, so that $x_1 = x_1^2$ by right cancellation. Again by right cancellation $x_2 x_1 = x_2 x_1^2$ implies $x_2 = x_2 x_1$. Therefore x_1 is the identity. Thus S is an infinite cyclic semigroup adjoined with identity. If S contains only one generator, clearly S is an infinite cyclic semigroup.

COROLLARY 3.16. *A finitely generated right n.t.o. semigroup S is o-Archimedean iff* $S = \{x, x^2, \ldots\}$ *for some* $x \in S$ *with* $x^n \leq x^{n+1}$ *for every natural number n.*

COROLLARY 3.17. *Let S be a finitely generated right and left n.t.o. semigroup. If* $x_1, x_2, \ldots, x_n$ *is the minimal set of generators, then every element of S is of the form* x_r^m, *where* $r = 1,2,\ldots,n$ *and m is any natural number. S is ordered by:*
$x_1 < x_1^2 \leq \ldots < x_2 \leq x_2^2 \leq \ldots < x_n \leq x_n^2 \leq \ldots$. *Furthermore S is commutative.*

<u>*PROOF*</u>. By 3.14, $x_i x_j = x_j$ for $i < j$. The analogous condition corresponding to the left n.t.o. property is $x_j x_i = x_j$ for $i < j$. Then $x_i x_j = x_j x_i = x_j$ for $i \quad j$. Hence the conclusion is evident from 3.14.

The following theorem about the location of generators is proved with the help of Pastjin while he was visiting Bowling Green.

THEOREM 3.18. *Let S be a right n.t.o. semigroup which is finitely generated by periodic elements. Then, among the minimal set of generators, each one of them except the largest of them is either an idempotent or is the immediate succeeding element of an idempotent and if the cardinal number of the minimal set of generators is r, then S has exactly r idempotents and its structure is described as in 3.14.*

<u>*PROOF*</u>. Let $\{x_1, x_2, \ldots, x_r\}$ be the minimal set of generators with $x_1 < x_2 < \ldots < x_r$. As in the proof of 3.14, every element of S is of the form $x_r^{n_r} x_{r-1}^{n_{r-1}} \ldots x_1^{n_1}$ where each n_i is a nonnegative integer; $x_j = x_i x_j$, $x_j = x_i^n x_j$ and $x_j > x_i^n$ for every natural number n and for $i < j$. Let the periods of x_j's be m_j's for $1 \leq j \leq r$. Now, if y is a product of x_j's with $j < i$, then

$x_i^{m_i} y = y x_i^{m_i} = x_i^{m_i}$. For, if t is the largest suffix of x_j occurring in the expression of y, then $y \leq x_t^N$ for some natural number N, which is less than x_i from above. So $x_i^{m_i} y \leq x_i^{m_i+1} = x_i^{m_i} \leq x_i^{m_i} y$. Thus $x_i^{m_i} y = x_i^{m_i}$. Similarly $y x_i^{m_i} = x_i^{m_i}$. We will use this fact in proving that there are no elements strictly in between $x_j^{m_j}$ and x_{j+1}. Suppose there exists an element y strictly in between $x_j^{m_j}$ and x_{j+1}. Then y is a product of x_t's, where $t < j + 1$. One of these t's is j since otherwise $y = x_{n_1}^{\alpha_1} \ldots x_{n_s}^{\alpha_s}$, where $n_i < j$ and $n_1 > n_2 > \ldots > n_s$. So, $y \leq x_{j-1}^{n}$ for some natural number n and hence in turn is strictly less than x_j, which is impossible. The exponent of x_j occurring in y is m_j, since otherwise if the exponent of x_j is t and if $t < m_j$, then $x_j^{m_j} \leq y \leq x_j^{t+1} \leq x_j^{m_j}$, which implies $y = x_j^{m_j}$. This is impossible. Since $x_j^{m_j} = x_j^{m_j+n}$ for every nonnegative integer n, the exponent of x_j is exactly m_j. Therefore $y = x_j^{m_j}$ from the previous observation since $x_j^{m_j}$ is the first factor of y starting from left. Next we claim that the only idempotents in S are $x_j^{m_j}$, $i \leq j \leq r$, which proves that every generator is an idempotent or is the immediate succeeding element of an idempotent from the previous observation and also S contains exactly r idempotents. Suppose that e is an idempotent such that $x_j < e < x_{j+1}$. Since there are no elements strictly in between $x_j^{m_j}$ and x_{j+1}, we must have $x_j < e \leq x_j^{m_j}$. By right n.t.o. condition, $e = x_j t$ for some $t \in S$. Therefore

$$e = e^{m_j} = (x_j t)(x_j t) \ldots m_j \text{ times} \geq x_j^{m_j} \geq e$$

and hence $e = x_j^{m_j}$. Suppose there exists an idempotent e and that $e > x_r$ and $e \neq x_r^{m_r}$. Clearly $x_r^{m_r}$ is the maximal element in S. So we must have $x_r \leq e < x_r^{m_r}$. As above it can be shown that if $e \neq x_r$, $e = x_r^{m_r}$.

From 3.17 and 3.18 the following are evident.

COROLLARY 3.19. *Let S be a right and left n.t.o. semigroup which is finitely generated by periodic elements. If* $\{x_1, x_2, \ldots, x_n\}$

is the minimal set of generators, then every element of S is of the form $x_i^{n_i}$, $1 \leq n_i \leq m_i$, *where* m_i *is the period of* x_i. *S is ordered by:*

$$x_1 < x_1^2 < \ldots < x_1^{m_1} < x_2 < x_2^2 < \ldots < x_i^{m_i} < x_{i+1} < \ldots$$

COROLLARY 3.20. *Let S be a finite right and left n.t.o. semigroup. Then there exist a finite number of elements* $x_1,\ldots,x_r$ *such that S is of the form described in the Corollary 3.19.*

Historically the structure of n.t.o. semigroups has attracted the attention of Clifford and Hölder because of the problem of determination of t.o. semigroups which are o-isomorphic with an additive subsemigroup of positive real numbers with the usual order. In fact we observe that one-sided n.t.o. condition is sufficient to obtain these results. The first step in their proofs is to establish a theorem of commutativity. The following is proved by Fuchs and Holder for two-sided n.t.o. case.

THEOREM 3.21. *Let S be an o-Archimedean right n.t.o. semigroup. Then S is commutative.*

PROOF. Since the only unit in positively ordered semigroups is the identity by (iii) of 2.1, S\1 is itself an o-Archimedean n.t.o. semigroup, if the identity 1 exists. So without loss of generality we can assume that S does not contain identity. If S contains a minimal element, then S is a cyclic semigroup by 3.2 and hence S is commutative. Suppose now that S has no minimal element. Let there exist a,b in S such that $ab \neq ba$. We shall show that $ba < ab$ is inadmissible and hence by symmetry $ab < ba$ is also inadmissible. Then we can conclude that S is commutative. If $ba < ab$, then $ab = bax$ for some x in S. Since S has no minimal element, there exists an y in S such that $y < x$, which implies $x = yy_0$ for some y_0

in S. Setting $z = \min(y,y_0)$, we have $z^2 \leq x$. We may assume that there exists an z such that $z^2 \leq x$, $z \leq a$, and $z \leq b$. If $z \leq a < b$, the above z does satisfy the properties. If $z \leq a$ and $b < a$, $z_1 = \min(b,z)$ does have the above properties. If $z > a$, and $a < b$, pick z_1 which is less than a and if $z > a$ and $a > b$, pick z_1 less than b. Then z_1 has the prescribed properties. Now by o-Archimedean property, $z \leq a$, and $z \leq b$ imply $z^m \leq a < z^{m+1}$, and $z^n \leq b < z^{n+1}$ for some natural numbers m and n. Then $ab = bax \geq z^n \cdot z^m \cdot z^2 \geq ab$ and so $ab = z^{m+n+2} = atz^{n+1}$, where $z^{m+1} = at$ for some t in S by right n.t.o. condition. If $b = tz^{n+1}$, then $b \geq z^{n+1}$. This contradicts the property of b. So ab is the maximal element of S by (vii) of 3.1. Suppose now $a < b$. Then $a^n < b \leq a^{n+1}$ for some $n \geq 1$. Therefore $b = a^n c$ for some $c \leq b$. Then $ca \leq ba < ab$ and so ca is not a maximal element. This implies $ac < ca$ is inadmissible as shown above. Similarly if ac is not a maximal element, then $ca < ac$ is inadmissible. In this case we must have $ac = ca$, which implies $ab = a^{n+1}c = (a^n c)a = ba$, which is a contradiction. If ac is a maximal element, then $ac = b$ and thus $b = ab = ba$ by positive order, which is again a contradiction. Hence S is commutative.

The commutative semigroup generated by the symbols x and y such that $xy = y$ and ordered by: $x < x^2 < \ldots < y < y^2 \ldots$, is a right n.t.o. semigroup which is not o-Archimedean. Hence o-Archimedean condition is not necessary for the commutativity of right n.t.o. semigroups. By virtue of (vi) of 2.1, we also observe that the positively t.o. idempotent semigroups are always commutative, without being o-Archimedean. We do not know whether commutative property is the necessary and sufficient condition for a right n.t.o. semigroup to be left n.t.o. However, as a consequence of 3.14, finitely generated right n.t.o. semigroups are left n.t.o. iff they are commutative.

Now we turn our attention to proving a classical result due to Hölder and Clifford. Here we adopt the proof given by Fuchs, because of its simplicity.

THEOREM 3.22. *Let S be an o-Archimedean right n.t.o. semigroup. Then S is o-isomorphic with a subsemigroup of one of the following t.o. semigroups:*

P: the additive semigroup of all nonnegative real numbers;

P_1: the real numbers in [0,1] with the usual ordering and with the multiplication $ab = \min(a + b, 1)$;

P_2: the real numbers in [0,1] and the symbol ∞ with the usual ordering and the operation $ab = a + b$ if $a + b \leq 1$ and $ab = \infty$ if $a + b > 1$. The first case occurs iff S is cancellative.

PROOF. By virtue of 3.21, S is commutative. As noted in the proof of 3.21, we may assume that S does not contain identity. If S contains a minimal element, then by 3.2, S is a cyclic semigroup and hence is of the form $\{a, a^2, \ldots\}$ with $a \leq a^2 \leq \ldots$. If no two powers of a are the same, then S is o-isomorphic with $\{1,2,3,\ldots\}$ which is a subsemigroup of P. If $a^n = a^{n+1}$, then the map $a^m \to m/n$ embeds S o-isomorphically in P_1. Suppose now that S has no minimal elements. Then,

(*) for every $x \in S$ and for every natural number t, there exists an z such that $z^t \leq x$.

For, in the proof of 3.21, we have seen that there exists an z such that $z^2 \leq x$. By the same reason there exists an z_1 such that $z_1{}^2 \leq z$ and hence $z_1{}^4 \leq x$. By induction, (*) is evident.

If S has a maximal element u, choose an $a < u$. If S has no maximal element, choose any $a \in S$. For every $b < u$, form two classes of paris of positive integers:

$$L_b = \{(m,n)\colon a \leq x^n \text{ and } x^m \leq b \text{ for some } x \in S\}$$

$$U_b = \{(k,\ell)\colon b \leq y^k \text{ and } y^\ell \leq a \text{ for some } y \in S\}$$

L_b is not empty since by (*), for every natural number m there exists an x such that $x^m \leq b$ and by o-Archimedean property $x^n \geq a$ for some natural number n. Similarly U_b is not empty. We claim now $m/n \leq k/\ell$ for all $(m,n) \in L_b$ and $(k,\ell) \in U_b$. For, by (*), for sufficiently large t we can choose $z \in S$ with $z^t \leq \min(x,y)$. Hence for $r \geq t$ and $s \geq t$ we can have $z^r \leq x < z^{r+1}$ and $z^s \leq y < z^{s+1}$. Therefore

$$z^{rm} \leq x^m \leq b \leq y^k \leq y^{(s+1)k} \text{ and similarly } z^{s\ell} \leq a \leq z^{(r+1)n}$$

This implies $rm \leq (s + 1)k$ and $s\ell \leq (r + 1)n$. It follows then $m/n \leq (1 + 1/r)(1 + 1/s)k/\ell$ for arbitrary large r and s and hence $m/n \leq k/\ell$. For sufficiently large t, we can find an z in S by (*), such that $z^t \leq \min(a,b)$. Then for $r_1 s \geq t$, we have $z^r \leq a < z^{r+1}$ and $z^s \leq b < z^{s+1}$. Then $s/r+1 \in L_b$ and $r/s+1 \in U_b$ with $s+1/r - s/r+1 \to 0$ as $t \to \infty$. Hence there exists one and only one real number β such that $m/n \leq \beta \leq k/\ell$ for all $(m,n) \in L_b$ and $(k,\ell) \in U_b$. We define now a mapping f from $S \backslash u$ into the set of all real numbers by setting $f(b) = \beta$. Clearly $f(a) = 1$. Also $f(b) > 0$ for all $b \in S \backslash u$. Since $b < c$ implies $L_b \subset L_c$, we have $f(b) \leq f(c)$. Moreover $f(bc) = f(b) + f(c)$ whenever $bc < u$, which is proved by using sufficiently small elements. $f(b) = f(c)$ iff $b = c$ since if $b < c$, then $c = bd$ for some $d \in S$, so that $f(c) = f(b) + f(d)$ and $f(d) = 0$, which is not true.

If S is cancellative, then it contains no maximal element and so f is an o-isomorphism of S into P. If S is not cancellative, then by (vii) of 3.1, S contains a maximal element, say, u and so $a^k = u$. If $b \neq u$, $b \leq u \leq a^k$, so that $f(b) \leq kf(a) = k$. Let $\alpha = g\ell b\{f(b): b(\neq u) \text{ in } S\}$. If there exists no c such that $f(c) = \alpha$, set $f(u) = \alpha$. Otherwise set $f(u) = \infty$, with the usual convention. The function $g(b) = (1/\alpha)f(b)$ for every $b \in S$, is obviously an o-isomorphism of S into P_1 and P_2. The fact that the first case occurs iff S is cancellative, is evident.

Naturally one may ask what algebraic properties imply o-Archimedean property and whether the o-Archimedean property is completely determined by algebraic properties. For right n.t.o. semigroups affirmative answers can be found.

DEFINITION. A semigroup S is called Archimedean if for every x and y in S there exists an integer n such that $x^n \in SyS$.

THEOREM 3.23. *Let S be a right n.t.o. semigroup not containing identity. Then the following are equivalent:*

(i) S is o-Archimedean;

(ii) S is Archimedean;

(iii) S is ζ-indecomposable.

PROOF. (i) ⟹ (ii). Suppose that there exist x and y in S such that $x^n \notin SyS$ for every natural number n. Then $x^n \notin yS^1$ for every n, since $x^n \in yS^1$ for some n implies $x^{n+2} \in SyS$. Then by (iii) of 3.1 $yS^1 \subset x^nS^1$ for every n. Thus $y = x^nz_n$ for every n and $y \geq x^n$. Hence $x^n = y$ for some n since S is o-Archimedean. Consequently $y \in yS$. Then S is a nilsemigroup by 2.5. Now it is clear that S is Archimedean.

(ii) ⟹ (iii). Let $P(\neq S)$ be a completely prime ideal in S. Then for $x \in S\backslash P$ and $y \in P$, we have $x^n \in SyS \subset P$, which implies $x \in P$. Hence S is ζ-indecomposable.

(iii) ⟹ (i) is evident by 2.9.

NOTE. Theorem 3.22 asserts that right n.t.o. semigroups which are o-Archimedean are either o-isomorphic with subsemigroups of the additive semigroup of positive real numbers under the usual order or nilsemigroups with 0 as a maximal element. Using this fact and Theorem 3.23, we can prove in the following that right n.t.o. semigroups satisfying certain algebraic conditions are o-isomorphic with subsemigroups of the additive semigroup of positive real numbers.

THEOREM 3.24. *Let S be a right Noetherian right n.t.o. semigroup in which $a \neq ba$ for every a and b in S. Then S is an infinite cyclic semigroup with the powers of the generator increasing in order.*

PROOF. Since S is right Noetherian, we can write $S = \bigcup_{i=1}^{n} x_i S^1$ with $x_1 < x_2 < \ldots < x_n$. Since right ideals are linearly ordered, $S = x_1 S^1$. Suppose that there exists an element $s \in S$, which is not a power of x_1. Then $s = x_1 s_1$ with $s_1 = x_1 s_2$. Continuing in this manner, we have $s_i = x_1 s_{i+1}$ for all $i = 1,2,\ldots$. By positive order, $s_{i+1} \leq s_i$ and hence $s_i S^1 \leq s_{i+1} S^1$. Thus we have an ascending chain of right ideals $\{s_i S^1\}$. By right Noetherian condition this implies $s_i S^1 = s_{i+1} S^1$ for some i. Therefore $s_i = s_{i+1} t$ and $s_{i+1} = s_i n$ for some $t, n \in S$. This implies that $s_{i+1} \leq s_i \leq s_{i+1}$ by positive order and hence $s_i = s_{i+1}$. Now $s_1 = x_1^{i+1} s_{i+1} = x_1(x_1^{i} s_{i+1}) = x_1 s$. This contradicts our hypothesis that $a \neq ba$ for any a and b in S.

For right cancellative semigroups without idempotents, the condition that $a \neq ba$ for any pair of elements a and b, is trivially true. In general the converse is not true. But in the above theorem we have proved the converse for right Noetherian, right n.t.o. semigroups.

LEMMA 3.25. *Let S be a right cancellative, right n.t.o. semigroup. If S contains an idempotent e, then e is the identity of S.*

PROOF. Since $xe = xe^2$ for every x in S, $x = xe$ by right cancellation. This implies $x \geq e$ by positive order. Now if $e \neq x$, $e < x$ implies $x \in eS$ by right n.t.o. condition and hence $x = ex$. Thus e is the identity of S.

THEOREM 3.26. *Let S be a right cancellative, right Noetherian and right n.t.o. semigroup. Then S is o-isomorphic with infinite cyclic semigroup* $\langle x\rangle$ *with* $x < x^2 < \ldots$, *possibly adjoined with identity.*

PROOF. If S has no idempotents, then $a = ba$ for some a and b implies that $ba = b^2a$ and hence $b = b^2$ by right cancellation. Hence $a \neq ba$ for any a and b in S and therefore in this case the result is evident from 3.24. Suppose that e is an idempotent of S. Then $e = 1$ by 3.25. Clearly $S\backslash e$ is a right Noetherian right cancellative right n.t.o. semigroup without idempotents and hence S is described as above.

We note that in the above theorems we have used only the ascending chain condition on principal right ideals and hence a weaker form of right Noetherian condition is sufficient for the validity of Theorem 3.24. Without the Noetherian condition one may obtain the same characterization if the semigroup satisfies an intersection property, which is stronger than the condition $a \neq ba$ for any pair of elements in the semigroup.

THEOREM 3.27. *Let S be a right n.t.o. semigroup in which* $\bigcap_{n=1}^{\infty} x^n S = \emptyset$ *for every* $x \in S$. *Thus S is o-isomorphic with a subsemigroup of the additive semigroup of positive real numbers under usual order.*

PROOF. Suppose that P is a completely prime ideal and $P \neq S$. Then there exists an $x \in S\backslash P$ and $x^n \notin P$ for every natural number n. Therefore $P \subset \bigcap_{n=1}^{\infty} x^n S^1$ by (iii) of 3.1. Thus $\bigcap_{n=1}^{\infty} x^n S^1$ is nonempty. This is impossible since $y \in \bigcap_{n=1}^{\infty} x^n S^1$ implies $xy \in \bigcap_{n=1}^{\infty} x^n S = \emptyset$. Thus S has no completely prime ideals. Hence, as indicated in the proof of Theorem 3.24, the conclusion is evident by noting that S cannot be a nilsemigroup.

THEOREM 3.28. *Let S be a right cancellative, right n.t.o. semigroup containing no idempotents. If* $|S| > 1$ *and if every subset in S has a minimal element, then S is o-isomorphic with an infinite cyclic semigroup, the powers of the generators increasing in order.*

PROOF. Suppose there exist x and y in S such that $x^n < y$ for every natural number n. Then $y = x^n s_n$. If $s_i \leq s_{i+1}$, then $x^{i+1}s_i \leq x^{i+1}s_{i+1}$, so that $xy \leq y \leq xy$, i.e., $xy = y$. This implies $x^2y = xy$ and hence $x^2 = x$, which is not true by hypothesis. Therefore we have a strictly decreasing sequence of elements, namely, $s_1 > s_2 > \dots$. This contradiction leads to the fact that S is o-Archimedean. Clearly by hypothesis S contains minimal elements. Hence by 3.2, the conclusion is evident.

A subset T of a t.o. semigroup S is *bounded above* if there exists an element $x \in S$ such that $t \leq x$ for every $t \in T$. x is called a *least upper bound* (ℓ.u.b.) of a subset T of a t.o. semigroup S if $t \leq x$ for every $t \in T$ and whenever $y < x$, then there exists an element $t \in T$ such that $y < t \leq x$. A t.o. semigroup is called *complete* if every subset, which is bounded above, has a ℓ.u.b.

Since the subsemigroups of the additive semigroup of positive real number under usual order satisfy the completion property, it is interesting to know what complete t.o. semigroups are subsemigroups of the positive real numbers. We begin with a few lemmas, which are interesting in their own right, for proving our main theorem.

LEMMA 3.29. *Let S be a complete right n.t.o. semigroup. If* $a^n < b$ *for every natural number n, then* $b = ab$.

PROOF. If a is periodic, then a^n is an idempotent for some natural number n and $a^n = a^{n+1}$ by positive order. Since S is right n.t.o., $a^n < b$ implies $b \in a^nS$ and so $b = a^nb = a \cdot a^n \cdot b = ab$.

Suppose that a is not periodic. Since the set of powers of a is bounded by b, this set has a ℓ.u.b. c since S is complete. If $a^n = c$ for some natural number n, then $a^{n+1} \leq c = a^n$ and so $a^{n+1} = a^n$, which contradicts that a is not periodic. So $a^n < c$ for every natural number n. In particular $a < c$, which implies by right n.t.o. condition, $c = ad$ for some d in c. Therefore $c \geq d$. Suppose $c > d$. Since c is the ℓ.u.b., we have then $d < a^n < c$ for some natural number n. Hence $a^{n+1} \leq c = ad \leq a^{n+1}$ and so $c = a^{n+1}$, which is not true as above. Thus $d = c$ and $c = ac$. If $c = b$, clearly $b = ab$. If $c < b$, then $b = ct$ for some $t \in S$ and so $b = ct = act = ab$.

LEMMA 3.30. *Right cancellative right n.t.o. complete semigroups contain only one non-o-Archimedean element, which is the identity.*

PROOF. Suppose that a is a non-o-Archimedean element and so there exists an element b such that $a^n < b$ for every natural number n. Then, by 3.29, $b = ab$, which implies $a = a^2$ by right cancellation. Thus $a = 1$ by 3.25.

THEOREM 3.31. *Let S be a right n.t.o. semigroup. Then S is o-isomorphic with a subsemigroup of the additive semigroup of positive real numbers with usual order iff S is complete and right cancellative.*

PROOF. Let S be right cancellative, right n.t.o. complete semigroup. If S does not contain idempotents, then S is o-Archimedean. If S contains an idempotent, which is the identity by 3.25, then, by 3.30, S is o-Archimedean right n.t.o. semigroup with identity. Hence S is a subsemigroup of additive semigroup of positive real numbers by 3.22.

THEOREM 3.32. *Let S be a right n.t.o. semigroup without identity. Then the following are equivalent:*

(i) S is a subsemigroup of the additive semigroup of positive real numbers with usual order;

(ii) S is a completely right cancellative semigroup;

(iii) S is a complete semigroup with $a \notin Sa$ for every $a \in S$;

(iv) S is a complete semigroup with $\bigcap_{n=1}^{\infty} a^n S = \emptyset$ for every $a \in S$.

<u>*PROOF*</u>. Clearly (i) implies all the other conditions. By 3.31, (ii) $\Rightarrow$ (i).

(iii) $\Rightarrow$ (i). If $a^n < b$ for some $a,b \in S$ and for some natural number n, then by 3.29, $b = ab$. This is not true. Hence S is o-Archimedean and thus by 3.22, the conclusion follows.

(iv) $\Rightarrow$ (i). As above if a is non-o-Archimedean, $b = ab$ for some $b \in S$. Hence $b = \bigcap_{n=1}^{\infty} a^n S = \emptyset$. Then (i) is evident from the previous observation.

We have discussed so far whether a right n.t.o. semigroup is a subsemigroup of the additive semigroup of positive real numbers if it is completely or o-Archimedean. Since subsemigroups of additive semigroups of positive real numbers are also ordinally irreducible (which we have introduced in Chapter II) we may now enquire the converse. The commutative version of the following results are due to Clifford [2] and [7]. In 2.15, we have shown that every positively t.o. semigroup is an ordinal sum of ordinally irreducible semigroups. Using this theorem, one can reduce the study of right n.t.o. semigroups to that of ordinally irreducible right n.t.o. semigroups.

THEOREM 3.33. *An ordinal sum $S = \cup S_\lambda$ of semigroups $\{S_\lambda\}$ is a right n.t.o. semigroup iff every S_λ is a right n.t.o. semigroup.*

PROOF. Let S be a right n.t.o. semigroup. If $x,y \in S_\lambda$ and if $x < y$, then $y = xt$. If $t \in S_\mu$ such that $\lambda < \mu$, then $xt = t$ and so $y \in S_\lambda \cap S_\mu = \emptyset$. If $t \in S_\mu$ with $\mu < \lambda$, then $xt = x$ and so $x = y$, which is a contradiction. Hence $\lambda = \mu$ and so S_λ is a right n.t.o. semigroup. Conversely assume that each S_λ is right n.t.o. Suppose $a < b$ with $a,b \in S$. If $a,b \in S_\lambda$, then clearly $b \in aS_\lambda \subset aS$. Let $a \in S_\lambda$, $b \in S_\mu$, and $\lambda \neq \mu$. Then $ab = b$. Thus S is a right n.t.o. semigroup.

DEFINITION. An ideal A in a semigroup S is *absorbent* if $ab = ba = a$ for every $a \in A$ and $b \in S \backslash A$.

THEOREM 3.34. *A right n.t.o. semigroup S is ordinally irreducible iff S contains no nonempty proper absorbent completely prime ideals.*

PROOF. Let S be ordinally irreducible. Suppose that P is an absorbent completely prime ideal. If $a \in S\backslash P$ and $b \in P$, then $b < a$ implies $a \in bS \subset P$. Hence $a < b$. Also, since P is an absorbent ideal, $ab = ba = b$. Thus S is the ordinal sum of the t.o. semigroups $S\backslash P$ and P in that order. This contradiction asserts that S has no nonempty proper absorbent completely prime ideals. Conversely let S have no absorbent nonempty proper completely prime ideals. If possible let $S = \cup S_\lambda$, where $\lambda \in \Lambda$, $|\Lambda| > 1$, Λ totally ordered set and each S_λ is a subsemigroup such that $\lambda < \mu$, $a \in S_\lambda$ and $b \in S_\mu$ implies $a < b$ and $ab = ba = b$. Since $|\Lambda| > 1$, there exist $\lambda,\mu \in \Lambda$ such that $\lambda < \mu$. Then $R = \{\alpha \in \Lambda : \alpha > \lambda \mid$ is a nonempty subset of $\Lambda\}$. Let $P = \bigcup_{\alpha \in R} S_\alpha$. Let $a,b \in P$. If a and b belong to the same S_λ then $ab \in S_\lambda \subset P$. If $a \in S_\alpha$ and $b \in S_\beta$, then $ab = a \in S_\alpha \subset P$ or $ab = b \in S_\beta \subset P$ according as $\alpha > \beta$ or $\beta > \alpha$. Thus P is a subsemigroup. If $x \in P$ and $y \notin P$, then $x \in S_\gamma$ and $y \in S_\delta$ where $\delta \notin R$ and $\gamma \in R$. Then $\gamma > \delta$ by the definition of R and so $xy = yx = x \in P$. Hence P is an absorbent ideal. We shall show now that P is completely prime. For, let $ab \in P$ for some a and b in S. Then $a \in S_\alpha$ and $b \in S_\beta$. If $a \neq b$ we may assume $\alpha > \beta$. Therefore $a = ab \in P$. Similarly if $\beta > \alpha$, $b = ab \in P$. If $a = b$, $a^2 \in S_\gamma$ for some $\gamma \in R$. If $a \notin S_\gamma$, $a \in S_\alpha$ where $\alpha \neq \gamma$. Then $a^2 \in S_\alpha$ since S_α is a subsemigroup. Therefore $\alpha = \gamma$, which is a contradiction. Since R is nonempty, P is nonempty. By assumption we have then $P = S$, i.e., $\bigcup_{\alpha \in R} S_\alpha = \bigcup_{\alpha \in \Lambda} S_\alpha$, so that $R = \Lambda$. This means $\lambda > \lambda$, which is a contradiction.

THEOREM 3.35. *Let S be a n.t.o. semigroup containing a nonzero idempotent which is not a minimal element. Then S is not ordinally irreducible.*

PROOF. Let $P = \{x \in S: x > e\}$ where e is nonzero idempotent. P is clearly a proper completely prime ideal. If $x \in S \backslash P$ and $y \in P$, then $x \leq e < y$, so that $xy \leq ey$ and $y = ey$ by n.t.o. condition. Therefore $xy \leq y$ and hence $xy = y$ by positive order. Thus P is an absorbent proper prime ideal. Then by 3.34, the conclusion follows.

THEOREM 3.36. *An ordinally irreducible, right and left n.t.o. semigroup S can contain at most one idempotent element, which, if it exists, is a zero of S.*

PROOF. Let $P = \{x \in S: x > e\}$, where e is an idempotent. It can easily be verified that P is a completely prime ideal. Let $x \in S \backslash P$ and $y \in P$. Then $x \leq e$ and hence $xy \leq ey$ and $yx \leq ye$. Since $y \in P$, $e < y$ and so $y \in eS \cap Se$ by right and left n.t.o. condition. Hence $y = ey = ye$. Thus $xy \leq y$ and $yx \leq y$, which implies, by positive order, $xy = y = yx$. Therefore P is absorbent. Then, by 3.34, $P = \emptyset$ since $P = S$ implies $e > e$. Hence e is the zero of S. Since the zero, if exists, is unique, S can have at most one idempotent.

One big class of ordinally irreducible semigroups is the class of all positively t.o. nilsemigroups since they are o-Archimedean and hence ordinally irreducible by 2.12. We shall now discuss some structural properties of right n.t.o. nilsemigroups.

THEOREM 3.37. *An ordinally irreducible, periodic, right and left n.t.o. semigroup is a nilsemigroup.*

PROOF. Since every element is periodic, some power of it is an idempotent, which is zero by 3.36. Hence the semigroup is nil.

THEOREM 3.38. *Let S be a right n.t.o. nilsemigroup. Then the following are equivalent:*

(i) S is a finite cyclic semigroup;
(ii) S is finitely generated;
(iii) S is finitely generated as a right ideal;
(iv) S contains a minimal element.

PROOF. Clearly (i) implies all the other conditions. Since (ii) and (iii) imply the existence of a minimal element, it is sufficient to prove (iv) implies (i). Let x be a minimal element of S. Suppose $y \in S$ and $y \neq x^n$ for any natural number n. Since $x < y$, we have $y = xt_1$ with $t_1 \neq x$. Again $t_1 > x$ and hence $t_1 = xt_2$ with $t_2 \neq x$. Continuing in this manner, we have $y = x^n t_n$ for every natural number n. Since $x^n = 0$ for some natural number n, $y = 0$ and thus the result is evident.

THEOREM 3.39. *Let S be a right n.t.o. nilsemigroup. Then the following are true:*

(i) If $a,b,c \in S$ and if $ab = ac \neq 0$, then $b = c$.
(ii) If $a,b \in S$ such that $a < b$ with $b \neq 0$, then $b = ac$ for a unique c in S.

PROOF. (i) Suppose $b < c$. Then $c = bd$ for some d in S. Now $ab = ac = abd$ and hence $ab = abd^n$ for every natural number n. Since some power of d is 0, ab = 0, which is not true. Thus b < c is inadmissible. Similarly c < b is inadmissible. Thus b = c.

(ii) Clearly b ε aS. Suppose b = ac = ad. Then by (i) c = d.

THEOREM 3.40. *Let S be a right n.t.o. nilsemigroup with no minimal element. Then the following are true:*

(i) If $a < b$ for some a,b in S with $b \neq 0$, then there exists an element $c \in S$ such that $a < c < b$.

(ii) $S = S^2$.

(iii) S contains no maximal one-sided or two-sided ideals.

(iv) S is not right or left Noetherian.

(v) S is not a finite semigroup.

PROOF. (i) Suppose that there exists no element x such that a < x < b. We can write b = ac for some c in S by right n.t.o. property. Since S has no minimal elements, there exists an d such that d < c. Then a $\leq$ ad $\leq$ ac = b. If a = ad, then a = 0 as observed in the proof of the above theorem. Hence 0 < b, which is impossible. Therefore a < ad $\leq$ b. Then by assumption, this implies ad = b = ac. Since d < c, we have c = de for some e in S. Hence ad = ade. Again by the above observation ad = 0. Therefore b = 0, which is a contradiction. By (v) and (vi) of 3.1, (ii)-(v) are evident.

For the complete description of some classes of ordinally irreducible semigroups, we shall now discuss some important facts of complete n.t.o. semigroups. Not every n.t.o. semigroup is complete because of Clifford's example in [8]. One may ask whether it is possible to embed a noncomplete n.t.o. semigroup in a complete n.t.o. semigroup. In order to settle this question Clifford [8] introduced the concept of normal completions. Krishnan [9] dealt with some general ideas. We now describe Clifford's method. If S and T are two t.o. semigroups and if S is isomorphic with a subsemigroup of T (and thus regarded as a subsemigroup of T) such that T is complete and every element of T is a least upper bound of some subset of S and also the greatest lower bound of some subset of S, then we say that T is a normal completion of S or T is embedded normally in S. Two normal completions T and L of S will be called equivalent if there is an isomorphism of T onto L leaving each element of S fixed. If S is an arbitrary totally ordered set, it possesses a normal completion which is unique to within equivalence, namely that constructed from Dedekind cuts. Hence consider the Dedekind completion Σ of a commutative t.o. semigroup S and try to extend the binary operation from S to all of Σ in every possible way. It can be proved that distinct ways of extending the operation lead to inequivalent normal completions. Whether there exists always at least one normal completion is an open question.

If S is a commutative t.o. semigroup, we can introduce order topology in S by taking the open intervals as a basis for the open sets of S. S is called lower semicontinuous if for every a,b,c in S such that $c < a + b$, there exists neighborhoods V_a of a and V_b of b such that $x \in V_a$ and $y \in V_b$ imply $c < x + y$. 'Upper semicontinuous' is defined dually. S is continuous if and only if it is both lower semicontinuous and upper semicontinuous.

It is proved by Clifford that if a commutative t.o. semigroup S is lower semicontinuous and also upper semicontinuous, then it has a unique lower semicontinuous normal completion and also has a unique upper semicontinuous normal completion. If these two normal completions coincide, then S admits a unique normal completion.

If T is a subset of an ordered set L, then T is said to be order-dense in L if (1) T contains the greatest and the least elements of L, if such exist, and (2) for every pair α,β of distinct elements of L, either there is an element of T between them, or else both α and β belong to T. It is easy to verify that if there is no element of T between α and β, then there is no element of L between them either. We say that an ordered set S is dense if $a < b$ (a,b in S) implies $a < x < b$ for some x in S.

Clifford proved that dense commutative n.t.o. semigroups are continuous and hence from the above we have that dense commutative n.t.o. semigroups have a unique normal completion, if one such exists.

The semigroups P, P_1, and P_2 mentioned in 3.22 together with the additive semigroup Z of positive integers and the additive semigroup $Z_n = \{0,1,2,\ldots,n - 1\}$ modulo n are called fundamental semigroups. These are all o-Archimedean and complete.

If a commutative n.t.o. semigroup S is noncancellative and is o-Archimedean not containing identity, then it is a nilsemigroup by 3.2. Suppose that S does not contain a minimal element. Then by Theorem 4 and its corollary in [7] and by noting nil totally ordered semigroups with 0 as a maximal element are ordinally irreducible, we assert that there exists a uniquely determined isomorphism of S onto an order-dense subsemigroup of either P_1 and P_2 (but not both) and P_1 and P_2 are, within isomorphism, the only complete commutative nilsemigroups which are n.t.o. If S contains a minimal element, then it is a finite cyclic semigroup by 3.38 and hence is complete.

If the commutative n.t.o. semigroup S is cancellative and o-Archimedean we have the following results due to Holder and Huntington respectively.

THEOREM 3.41. *Let S be a commutative n.t.o. semigroup satisfying:*

(1) the cancellation law holds in S,

(2) S has no identity element, and

(3) S has no minimal element.

Then S is isomorphic with P iff it is complete and S can be normally embedded in P iff it is o-Archimedean.

THEOREM 3.42. *Let S be a commutative n.t.o. semigroup satisfying:*

(1) the cancellation law holds in S,

(2) S has no identity element, and

(3) S has a least element.
Then if S is o-Archimedean, it is also complete, and is isomorphic with Z.

We proved in 2.12 that o-Archimedean positively t.o. semigroups without identity are ordinally irreducible. But ordinally irreducible positively t.o. semigroups, even if they satisfy the n.t.o. condition, need not be o-Archimedean. The following counter-example of Clifford proves this situation [8].

Let G be the group of all lattice points (x,y) in the plane, where x and y are integers, under addition and ordered lexicographically: $(x,y) < (x^1,y^1)$ if $x < x^1$ or if $x = x^1$ and $y < y^1$. Let G_+ be the set of all $(x,y) > (0,0)$. Define a binary relation ρ in G_+ as follows. Each of the elements (0,y) and (1,y) of G_+ bears the relation ρ only to itself. For x and $x' \geq 2$, $(x,y)\rho(x',y')$ if and only if $x = x'$. ρ is a convex congruence relation and so we can order G_+/ρ in an obvious way. If we write a = (0,1) and b = (1,0), and take nb = (n,0) as the representative of the congruence class of all (n,y), then $S = G_+/\rho$ can be expressed as an ordered set

$$\{a < 2a < 3a < \ldots < b-2a < b-a < b < b+a < b+2a < \ldots 2b < \ldots\}$$

with addition defined so that 2b + na = 2b for any integer n. S is a commutative ordinally irreducible n.t.o. semigroup and is not o-Archimedean, which can be observed in 3.43.

Essentially the following theorems are due to Clifford who assumed that the semigroups under consideration are commutative. Because of 3.22 we can as well start with the assumption of being noncommutative.

THEOREM 3.43. *Let S be a complete, ordinally irreducible right and left n.t.o. semigroup. Then S is o-isomorphic with one of the fundamental semigroups.*

PROOF. Clearly ordinally irreducible t.o. semigroups have no identity. Then by virtue of 3.41, 3.42 and the remarks preceding to these and 3.22, it suffices to prove that S is o-Archimedean. Let $a,b \in S$. If $a^n = a^{n+1}$ for some natural number n, then a^n is an idempotent, so that a^n is the maximal element by 3.36. Hence $a^n \geq b$. Suppose that a is not periodic. Assume, by way of contradiction, $a^n < b$ for every natural number n. Since the set $\{a^n: n \geq 1\}$ is bounded above, it has a least upper bound, say c. If $a^n = c$ for some natural number n, then $a^{n+1} < c = a^n$, which implies that a is periodic, contrary to the assumption. Therefore $a^n < c$ for every natural number n. Since $a < c$, $c = ad$ for some d in S because of right n.t.o. condition. Suppose $c > d$. Then, by the least upper bound property of c, $d < a^n$ for some natural number n and hence $c \leq a \cdot a^n = a^{n+1}$, contrary to what we have just shown. Hence $c = d$, and so $c = ac$. By using left n.t.o. condition, we can show in a similar fashion $c = ca$.

Let $S(< c) = \{y \in S: y < c\}$ and $S(\geq c) = \{y \in S: y \geq c\}$. $S(< c)$ is a subsemigroup of S. For if $x,y \in S(< c)$, then $x < a^n$ and $y < a^m$ for some natural numbers n and m, whence $xy \leq a^{n+m} < c$. $S(\geq c)$ is also a subsemigroup of S. For, if $x,y \in S(\geq c)$, then $x \geq c$, $y \geq c$ and so $xy \geq c^2 \geq c$, which implies $xy \in S(\geq c)$. Next we prove that if $x \in S(< c)$ and $y \in S(\geq c)$, then $xy = y = yx$. For, $x < a^n$ for some n and so $xc \leq a^n c = c$ since $c = ac$ and consequently $xc = c$. Now $y = c$ or $y = cd$ for some d in S, so that $xy = xc = c = y$ or $xy = (xc)d = cd = y$. Thus $xy = y$. Similarly $yx = y$. Hence S is the ordinal sum of the subsemigroups $S(< c)$ and $S(\geq c)$, contrary to the hypothesis that S is ordinally irreducible. Hence we conclude that $a^n \geq b$ for some n. Thus S is o-Archimedean.

By a cut in an ordered set S we mean a pair (L,R) of subsets L,R of S such that every element of L is less than every element of R and $S = L \cup R$. A subset A of S is called a lower (upper) class if $a \in A$ and $x < a$ $(x > a)$ imply $x \in A$. A is a lower class of S iff $B = S \setminus A$ is an upper class.

THEOREM 3.44. *Let S be a right and left n.t.o. semigroup, and let $S = \bigcup_{i \in I} S_i$ be its reduction into ordinally irreducible components S_i. Then S is complete if and only if the following conditions are satisfied:*

(1) The ordered set I is complete.

(2) Each S_i is isomorphic with a fundamental semigroup.

(3) If i is an element of I having no immediate successor, but is not the greatest element of I, then S_i must have a greatest element.

(4) If i is an element of I having no immediate predecessor, but not the least element of I, then S_i must have a least element.

(5) If $i < j$ is an adjacent pair of elements of I, then either S_i must have a greatest element or S_i must have a least element.

PROOF. If S is complete, then every S_i is complete and hence is isomorphic with a fundamental semigroup by 3.43. If I is not complete, then a gap in I, i.e., a nontrivial cut such that the lower (upper) class contains no greatest (least) element, would lead in an evident manner to a gap in S. Similarly, a gap in S would result if any one of the conditions (3), (4) or (5) is not valid.

Conversely let conditions (1)-(5) hold. Let A be a nonempty subset of S bounded from above. Let J be the set of all i in I such that $S_i \cap A \neq \emptyset$. J is bounded above. For, there exists an $t_\alpha \in S_\alpha$ such that $a < t_\alpha$ for every α. If $i \in J$, then there exists an $a_i \in S_i \cap A$, which implies $a_i < t_\alpha$ and so $i < \alpha$. Since I is complete, J has a least upper bound, say j. First suppose $j \in J$. Then $A_j = A \cap S_i$ is nonempty. If A_j is bounded in S_j, it has a least upper bound a, since by (2), S_i is complete; a is then evidently the least upper bound of A. If A_j is not bounded in S_j, S_j has no greatest element. If j has an immediate successor k, then by (5), S_k must have a least element a, and α is then the least upper bound of A. If j has no immediate successor, then by (3) j must be the greatest element of I; but this is impossible since it would imply A is not bounded in S. Next suppose $j \notin J$. j cannot have an immediate predecessor i, for then i would be an upper bound of J, yet $i < j$. j cannot be the least element of I, since this would imply that A is empty. By (4), we infer that S_j has a least element a, and a is evidently then the least upper bound of A.

Using this result Clifford obtains a necessary and sufficient condition for the embeddability of noncomplete t.o. semigroups in a complete t.o. semigroup [8].

THEOREM 3.45. *Let S be a commutative n.t.o. semigroup, and let* $S = \bigcup_{i \in I} S_i$ *be its reduction into ordinally irreducible components* S_i. *Then S can be embedded in a complete commutative n.t.o. semigroup iff each* S_i *is o-Archimedean. In this case, there exist one and, to within equivalence, only one normal completion of S.*

A pair (A,B) of subsets of a t.o. semigroup S will be called proximal if (1) $a \leq b$ for every a in A, b in B, and (2) there is at most one element c of S such that $a \leq c \leq b$ for every a in A, b in B.

THEOREM 3.46. *Let S be a continuous t.o. semigroup. Then S admits a continuous normal completion iff the following condition is satisfied: If (A,B) and (A',B') are proximal pairs of subsets of S, then (AA',BB') is a proximal pair.*

Recall, from Chapter 2, that for any positively ordered semigroup S, A denotes the set of all o-Archimedean elements of S and Q denotes the union of all proper completely prime ideals of S. We shall now discuss some descriptive properties of right n.t.o. semigroups.

THEOREM 3.47. *Let S be a right n.t.o. semigroup. Then the following are true:*

(i) $S = A \cup S\backslash A$, *where A is an o-Archimedean subsemigroup and $S\backslash A$ is a right n.t.o. semigroup.*

(ii) If x is a nonperiodic element in S and if $\bigcap_{n=1}^{\infty} x^n S$ *is nonempty, then* $\bigcap_{n=1}^{\infty} x^n S$ *is a completely prime ideal.*

(iii) If $A = \emptyset$, *then* $\bigcap_{n=1}^{\infty} x^n S \neq \emptyset$ *for every x.*

(iv) If $\bigcap_{n=1}^{\infty} x^n S \neq \emptyset$ *for every x in S and if no element of S is periodic, then* $A = \emptyset$.

(v) If S is right cancellative and if $A = \emptyset$, *S can have at most one idempotent and at most one periodic element.*

(vi) If $A \neq S$ *and if* $x \in A$, *then* $\bigcap_{n=1}^{\infty} x^n S = \emptyset$ *or* $\bigcap_{n=1}^{\infty} x^n S = 0$ *and* $\bigcap_{n=1}^{\infty} Sx^n = \emptyset$ *or* $\bigcap_{n=1}^{\infty} Sx^n = 0$.

(vii) If S is commutative, then $I = \{x \in S: \bigcap_{n=1}^{\infty} x^n S = \emptyset\}$ *is a completely prime ideal, if nonempty.*

(viii) If S is commutative and if A and I are nonempty and if $S \neq A$, *then* $A = I$.

(ix) If S is commutative, then $S = I \cup S\backslash I$, *where I is an o-Archimedean semigroup and $S\backslash I$ is a n.t.o. semigroup.*

(x) If $S \neq Q$, *then $S\backslash Q$ is an o-Archimedean right n.t.o. semigroup.*

(xi) $Z_r = \{a: ba = ca \text{ with } b \neq c\}$ *is a completely prime ideal and* $Z_r \subset \bigcap_{x \notin Z} xS$.

(xii) If $A \neq S$ *and if* $Z_r \neq \emptyset$, $A \not\subset Z_r$. *Always S is either right cancellative or S is o-homomorphic with a right cancellative right n.t.o. semigroup adjoined with 0.*

PROOF. (i) This readily follows from (x) of 3.1.

(ii) Let $ab \in \bigcap_{n=1}^{\infty} x^n S \neq \emptyset$. If $a,b \notin \bigcap_{n=1}^{\infty} x^n S$, then there exist natural numbers m_1 and m_2 such that $a \notin x^{m_1} S$ and $b \notin x^{m_2} S$, which implies, by right n.t.o. condition, $x^{m_1} \varepsilon aS^1$ and $x^{m_2} \varepsilon bS^1$. Hence $x^{m_1} \geq a$ and $x^{m_2} \geq b$ and so $x^{m_1+m_2} \geq ab$. But by hypothesis $ab \geq x^{m_1+m_2+1}$. Therefore $x^{m_1+m_2} = x^{m_1+m_2+1}$, which contradicts that x is not periodic. Hence $\bigcap_{n=1}^{\infty} x^n S$ is a completely prime ideal.

(iii) Let $x \varepsilon S$. Since $x \notin A$, there exists an y such that $x^n < y$ for every natural number n. Then by right n.t.o. condition, $y \varepsilon x^n S$ for every n. Thus $\bigcap_{n=1}^{\infty} x^n S \neq \emptyset$.

(iv) Let $x \varepsilon S$. There exists an y in $\bigcap_{n=1}^{\infty} x^n S$. By positive order $x^n \leq y$ for every n. If $x^n = y$ for some n, then $x^n \varepsilon x^{n+1}S$, which implies, by positive order, $x^n = x^{n+1}$ and hence x is periodic. This contradiction leads to the fact $x^n < y$ for every n. Hence $A = \emptyset$.

(v) Let e and f be idempotents in S such that $e < f$. Then by right n.t.o. condition $f = ef$, which implies $f = e$ by right cancellation. Now if x is a periodic element, then $x^m = x^{m+1}$ for some integer m and hence by right cancellation x is an idempotent, which is unique as above, if it exists.

(vi) Let $y \varepsilon \bigcap_{n=1}^{\infty} x^n S \neq \emptyset$. Then $y \geq x^n$ for every natural number. Since x is o-Archimedean, $y = x^n$ for some n. Then $x^n \varepsilon x^{n+1}S$ implies $x^n = x^{n+1}$ by positive order. Since x is o-Archimedean, the idempotent x^n is also o-Archimedean. Then, for some y, $x^n < y$ is impossible unless $x^n = 1$. So, either x^n is a maximal element of S or $x^n = 1$. But by (iii) of 2.1, $x^n = 1$ implies $x = 1$ and so $1 \varepsilon A$, which contradicts $A \neq S$. But by (i) of 2.1, x^n is a zero element of S. Then $\bigcap_{n=1}^{\infty} x^n S = 0$. The second part can be treated in a similar fashion.

(vii) Let I be nonempty. Suppose $x \varepsilon I$ and $s \varepsilon S$. If $y \varepsilon \bigcap_{n=1}^{\infty} (xs)^n S$ for every natural number n, then $y \varepsilon \bigcap_{n=1}^{\infty} x^n S = \emptyset$. Hence $xs \varepsilon I$ and so I is an ideal. To prove that I is completely prime, assume $ab \varepsilon I$ and $a,b \notin I$. Then there exist y and z such that $y \varepsilon \bigcap_{n=1}^{\infty} a^n S$ and $z \varepsilon \bigcap_{n=1}^{\infty} b^n S$. Hence $yz \varepsilon \bigcap_{n=1}^{\infty} (ab)^n S = \emptyset$. Hence I is completely prime.

(viii) By (x) of 3.1, $A \subset I$, since I is a completely prime ideal by (vii). Suppose now $x \in I \setminus A$. Then there exists an y such that $x^n < y$ for every natural number n. So $y \in x^n S$ for every n by right n.t.o. property and so $\bigcap_{n=1}^{\infty} x^n S \neq \emptyset$, which is absurd. Thus $A = I$.

(ix) This is an easy consequence of (viii).

(x) By 2.30, $S \setminus Q$ is o-Archimedean. It can easily be verified that S is a right n.t.o. semigroup.

(xi) Since $ba = ca$ implies $bas = cas$ for every s in S, Z_r is a right ideal and hence two-sided by (iv) of 3.1. Z_r is completely prime since, if $xy \in Z_r$ with $x,y \notin Z_r$, then we have $bxy = cxy$; $bx = cy$ and so $b = c$.

(xii) By (x) of 3.1, the first statement is obvious. If $Z_r = \emptyset$, then S is right cancellative. If $Z_r \neq \emptyset$, then by (xi) Z_r is a completely prime ideal and by (ii) of 3.1, Z_r is convex. So S/Z_r is a right n.t.o. right cancellative semigroup adjoined with zero. Then the canonical homomorphism $S \twoheadrightarrow S/Z_r$ proves the second statement.

We have noted before that the kernel of every positively t.o. semigroup S is either empty or 0. If kernel is empty, evidently S is a subdirect product. If the kernel is 0, then for every $x \neq 0$ there exists an ideal, say M_x, not containing x and so the intersection of all M_x's is zero. Thus S is again a subdirect product. Furthermore if S is a right n.t.o. semigroup, then S is a subdirect product of right n.t.o. subsubsemigroups since every ideal is convex by (ii) of 3.1 and so the Rees-factor semigroups are also right n.t.o. semigroups.

If a positively t.o. semigroup contains 0, then clearly every o-Archimedean element is nilpotent and conversely every nilpotent element is o-Archimedean. Hence the set N of all nilpotent elements, which is equal to A (the set of all o-Archimedean elements) is a completely prime ideal by 2.24. It can be verified easily that $N = A$ is a convex ideal. But if the positively t.o. semigroup is a right n.t.o. semigroup not containing 0, then also A is a convex ideal for if $x < y < z$ and $x,z \in A$, then y is a multiple of x and hence belongs to A.

Every right n.t.o. semigroup contains a unique maximal ideal, if it exists, since the right ideals are linearly ordered by set-inclusion and also it is convex. In fact this is true for globally idempotent positively t.o. semigroups.

PROPOSITION 3.48. *Let S be a positively t.o. semigroup with* $|S| > 1$. *If* $S = S^2$, *then every maximal ideal, if exists, is of the form* $S\backslash y$, *where y is a minimal element of S and hence S contains a unique maximal ideal, if it exists and it is convex.*

PROOF. Let M be a maximal ideal. Then by (iv) of 2.1, $S = M \cup y$, $y \notin M$, so that $S = S^2 = MS \cup yS = M \cup yS$. Now $y = ys$ for some $s \in S$. Since $y \notin M$, $s \notin M$ and so $y = y^2$. Clearly $y \neq yx$ for any $x \neq y$ in S. We claim that y is a minimal element. By way of contradiction, assume that $x < y$ for some $x \in S$. Then $y \leq yx \leq y^2 = y$ and so $y = yx$, which is a contradiction. The uniqueness and convexity of the maximal ideal is evident.

PROPOSITION 3.49. *Let S be a positively t.o. semigroup. If* $S \neq S^2$, *then the only maximal ideals are* $S\backslash x$, $x \in S\backslash S^2$ *and* $S\backslash y$, *where y is a minimal element of S, if it exists.*

PROOF. Clearly if $x \in S\backslash S^2$, then $S\backslash x$ is a maximal ideal. Suppose that M is a maximal ideal which is different from $S\backslash x$ where $x \notin S^2$. Then $M = S\backslash y$ for some y in S by (iv) by 2.1. Clearly $y \in S^2$. Since $S = M \cup y$, $S^2 = MS \cup yS \subset M \cup yS$, and so $y \in yS$. As in the proof of the above proposition, we assert that $y = y^2$ and y is a minimal element. Thus there exists only one M of this sort, if it exists.

If S is a nonglobally idempotent right n.t.o. semigroup, then $S\backslash S^2$ contains only a single element x and $S\backslash x$ is the only maximal ideal; which of course is convex. But if S is a nonglobally idempotent positively t.o. semigroup which is not necessarily a right n.t.o. semigroup, $S\backslash S^2$ may contain more than one element. If $S\backslash S^2$ contains only one element, say x, then S contains at most two maximal ideals, namely $S\backslash x$ and $S\backslash y$, where y is a minimal element. Though $S\backslash y$ is necessarily convex, $S\backslash x$ need not be convex.

Every right n.t.o. semigroup is combinatorial, which means that the only subgroups are one element groups. In fact this is the property of every positively t.o. semigroup. For, if G is a subgroup with an identity e, and if $x \in G$, we have $xy = e$ for some y in G. Hence $e \geq x$. But $x = ex$ implies $x \geq e$. Thus $x = e$.

In the previous chapter, we have noted that in a positively t.o. semigroup S, A (the set of all o-Archimedean elements) and $S\backslash Q$, where Q is the union of all proper completely prime ideals are convex subsemigroups. If S is a right n.t.o. semigroup, it can easily be verified that $Q\backslash A$ is also a convex subsemigroup. Thus S is a disjoint union of three convex subsemigroups, namely, A, $Q\backslash A$, and $S\backslash Q$, where A and $S\backslash Q$ are o-Archimedean. The question whether $Q\backslash A$ is also o-Archimedean remains open.

THEOREM 3.50. *Every right n.t.o. semigroup is a semilattice of convex subsemigroups, which are o-Archimedean positively t.o. semigroups.*

PROOF. By 2.10, if S is a right n.t.o. semigroup, it is a semilattice of its N-classes, which are o-Archimedean. We claim that every N-class T is convex. Let $x < y < z$ with $x, y \in T$. If x belongs to a completely prime ideal P, then $y \in P$ since $y \in xS \subset P$. Let $x \notin P$ and $y \in P$, where P is a completely prime ideal. Then $z \in yS \in P$. This implies $x \in P$ since x and z belong to the same completely prime ideal. This contradiction asserts that $y \in T$. Thus T is convex.

One may conjecture that every N-class of a right n.t.o. semigroup may be a right n.t.o. semigroup. The following example disproves this assertion.

Let $S = \{x < x^2 < \dots < y < yx < yx^2 < \dots < y^2 < y^2x < \dots\}$ with $y = yx$. The only proper completely prime ideal is $\{y, y^2, yx, \dots\}$ and so $\{y, y^2, yx, \dots\}$ is a N-class. This is not a right n.t.o. subsemigroup since $y < yx$ and $yx \neq yz$ for any z in this subsemigroup.

We shall now conclude this section with a construction of some right n.t.o. semigroups.

THEOREM 3.51. *Let N be the set of all nonnegative integers and G be a group. Suppose that I is a function from* $G \times G \to \{0,1\}$ *satisfying the following properties:*

(i) $I(\alpha,\beta) + I(\alpha\beta,\gamma) = I(\alpha,\beta\gamma) + I(\beta,\gamma)$ *for every* α, β, γ *in G;*

(ii) $I(\varepsilon,\varepsilon) = 1$, ε *being the identity of G;*

(iii) *There exists a nonperiodic element* $\alpha \in G$ *such that for all but a finite number of n,* $I(\alpha^n,\alpha) = 0$;

(iv) $I(\alpha,\alpha^{-1}) = 0$ *for every* $\alpha \neq \varepsilon$ *in G;*

(v) *For* $\alpha \neq \beta$, $\alpha \neq \varepsilon$, *and* $\beta \neq \varepsilon$, *one of* $I(\beta,\beta^{-1}\alpha)$ *and* $I(\alpha,\alpha^{-1}\beta)$ *is zero and the other is 1;*

(vi) *If* $I(\beta,\beta^{-1}\alpha) = 0$, *then for every* γ, *we have only the following possibilities:*

(a) $I(\beta,\gamma) = 0$ *and* $I(\alpha,\gamma) = 1$

(b) $I(\beta,\gamma) = I(\alpha,\gamma) = I(\beta\gamma,(\beta\gamma)^{-1}\alpha\gamma) = 0$

(c) $I(\beta,\gamma) = 1 = I(\alpha,\gamma)$ *and* $I(\beta\gamma,(\beta\gamma)^{-1}\alpha\gamma) = 0$

(vii) *If* $I(\beta,\beta^{-1}\alpha) = 0$, *then for every* γ, *we have only the following possibilities:*

(a) $I(\gamma,\beta) = 0$ *and* $I(\gamma,\alpha) = 1$

(b) $I(\gamma,\beta) = I(\gamma,\alpha) = I(\gamma\beta,(\gamma\beta)^{-1}\gamma\alpha) = 0$

(c) $I(\gamma,\beta) = 1 = I(\gamma,\alpha) = I(\gamma\beta,(\gamma\beta)^{-1}\gamma\alpha) = 0$

(viii) *If* $I(\beta,\gamma) = 1$ *and* $I(\alpha,\gamma) = 0$, *then* $I(\beta\gamma,(\beta\gamma)^{-1}\alpha\gamma) = 0$;

(ix) *If* $I(\gamma,\beta) = 1$ *and* $I(\gamma,\alpha) = 0$, *then* $I(\gamma\beta,(\gamma\beta)^{-1}\gamma\alpha) = 0$.

Define multiplication in $N \times G$ *by:*

$$(m,\alpha)(n,\beta) = (m + n + I(\alpha,\beta),\alpha\beta)$$

Define order in $N \times G$ *by: if* $m \geq 1$, $(m,\alpha) > (0,\beta)$; $(0,\varepsilon) > (0,\beta)$ *if* $\beta \neq \varepsilon$; *for* $\alpha \neq \beta$, $\alpha \neq \varepsilon$, $\beta \neq \varepsilon$, $(0,\alpha) > (0,\beta)$ *if* $I(\beta,\beta^{-1}\alpha) = 0$; *if* $m > n$, $(m,\alpha) > (n,\beta)$; $(m,\alpha) > (m,\beta)$ *if* $I(\beta,\beta^{-1}\alpha) = 0$. *Then* $N \times G$ *is a right n.t.o. semigroup not containing 1 and is not o-Archimedean but contains a central cancellable o-Archimedean element and infinite number of elements. Conversely every such semigroup can be constructed in this way.*

PROOF. By direct verification one can establish that $N \times G$ is a right n.t.o. semigroup not containing 1. By (iii) there exists an α such that all but a finite number, say m of $I(\alpha^i,\alpha)$ are zero. Then for any n, $(0,\alpha)^n = (t,\alpha^n)$ and $t < m + 1$ since α is not periodic and hence $(0,\alpha)^n < (m + 1,\varepsilon)$. Therefore $N \times G$ is not o-Archimedean. It is easy to verify that $(0,\varepsilon)$ is a central o-Archimedean element.

Conversely let S be a right n.t.o. semigroup containing a central cancellable o-Archimedean element a. Then by (viii) and (ix) of 3.1, P* contains a. By (iv) of 3.1 right ideals are two-sided. So by 2.42, S is isomorphic with $N \times G$ with the I-function satisfying the first two properties stated in the theorem. If $\alpha \neq \varepsilon$ and if $(0,\alpha)$ is in $(0,\varepsilon)S$, then we must have $(0,\alpha) = (0,\varepsilon)(0,\alpha)$, which implies $I(\varepsilon,\alpha) = 0$, which contradicts the consequence of the properties (i) and (ii), namely $I(\varepsilon,\alpha) = 1$. Since S is a right n.t.o. semigroup we must have then $(0,\varepsilon)$ is $(0,\alpha)S$, which implies $(0,\varepsilon) = (0,\alpha)(0,\alpha^{-1})$ and so $I(\alpha,\alpha^{-1}) = 0$. Now set $\gamma = \beta^{-1}$ in (i). Then, if $\beta \neq \varepsilon$,

$$I(\alpha,\beta) + I(\alpha\beta,\beta^{-1}) = I(\alpha,\varepsilon) + I(\beta,\beta^{-1}) = 1$$

Thus $I(\alpha,\beta) = 0$ or 1 and hence $I(\alpha,\beta) = 0$ or 1 for every α,β in G. Therefore I maps $G \times G$ into $\{0,1\}$.

To prove (iii), observe that S contains elements which are not o-Archimedean. Therefore there exist (m,α) and (n,β) in $N \times G$ such that $(m,\alpha)^r < (n,\beta)$ for every natural number r. If $m \neq 0$ and $n \neq 0$, then there exists an k such that $km > n$. Then

$$(m,\alpha)^{k+1} = [(k + 1)m + I(\alpha,\alpha) +\ldots+ I(\alpha^{k+1},\alpha),\alpha^{k+1}]$$

and hence $(m,\alpha)^{k+1} > (n,\beta)$, which is not true. If $m = 0$ and $n \neq 0$, then $(0,\alpha)^r < (n,\beta)$ for all r implies $\sum_{i=1}^{r} I(\alpha^i,\alpha) \leq n$ for every natural number r. Therefore all but a finite number of $I(\alpha^i,\alpha)$ are zero. If $m = n = 0$, then it is clear that $I(\alpha^i,\alpha) = 0$ for every i. The case $m \neq 0$ and $n = 0$ is inadmissible. α is not periodic since otherwise if $\alpha^s = \varepsilon$, then $(0,\alpha)^s = (N,\varepsilon)$ where $N = I(\alpha,\alpha) +\ldots+ I(\alpha^s,\alpha)$ and $(0,\alpha)^{ts} = [t(N + 1) - 1,\varepsilon]$. Choose k such that $k(N + 1) - 1 > n$. Therefore $(0,\alpha)^{ks} = [k(N + 1) - 1,\varepsilon] > (n,\varepsilon) > (n,\beta)$, which is not true.

It is routine to check that the remaining conditions are necessary for the admissibility of right n.t.o. structure on $N \times G$.

COROLLARY 3.52. *Let S be a commutative n.t.o. semigroup which is not o-Archimedean. If S does not contain identity and if S contains a cancellable o-Archimedean element, then S is isomorphic with $N \times G$ satisfying the properties stated in the above theorem together with: $I(\alpha,\beta) = I(\beta,\alpha)$ for every α,β in G.*

The commutative example supporting the hypothesis of the theorem is $N \times G$, where G is the additive group of integers and the I-function is defined by:

$$I(n,m) = 0 \text{ if } n,m > 0;\ I(n,-m) = 0 \text{ if } 0 < n \leq m$$

and in all other cases the value of I-function is 1.

Since $N \times G$ is right and left cancellative, the above right n.t.o. semigroup is in fact cancellative. By condition (iii) of the theorem the group G is necessarily nonperiodic. $N \times G$ is not finitely generated since cancellative finitely generated right n.t.o. semigroups are infinite cyclic semigroups possibly adjoined with 1 by virtue of 3.15. More than this we do not know the structure of G.

CHAPTER 4

ON ANOMALOUS PAIRS

We have noted in the previous chapter that Clifford and Hölder used n.t.o. condition for proving a t.o. semigroup to be o-isomorphic with a subsemigroup of the additive group of positive real numbers under usual order. But this is only a sufficient condition since the set of all natural numbers greater than or equal to 2, under addition and with the usual order is not a n.t.o. semigroup. Alimov observed that the absence of anomalous pairs is an intrinsic property of the additive semigroup of positive real numbers and he proved that this condition is necessary and sufficient for cancellative t.o. semigroups to be o-isomorphic with a subsemigroup of additive group of real numbers. We shall discuss this concept of anomalous pairs and its various implications.

According to Alimov two distinct elements a and b of a t.o. semigroup are said to form an *anomalous pair* if

$$a^n < b^{n+1} \text{ and } b^n < a^{n+1} \text{ for all natural numbers } n$$

or

$$a^n > b^{n+1} \text{ and } b^n > a^{n+1} \text{ for all natural numbers } n$$

Clearly the first alternative occurs if a and b are positive elements and the second if a and b are negative elements.

THEOREM 4.1. *Let S be a cancellative t.o. semigroup. If S has no anomalous pairs, then S is o-Archimedean. If S is o-Archimedean and right n.t.o., then it contains no anomalous pairs.*

PROOF. Let a and b be positive elements such that $a^n < b$ for every natural number n. Assume $a \neq 1$, if the identity 1 exists. We claim that for every pair of positive elements x,y in S, $y \neq 1$, $xy > x$ and $xy > y$. For, if $xy = x$, then $y = y^2$ by cancellation and hence $y = 1$. This is impossible. Similarly it can be proved that $xy > y$. Let $ab \leq ba$. By positive order, $b \leq ba$ and hence

$$b^n \leq (ba)^n < a(ba)^n < a(ba)^n b = (ab)^{n+1}$$

Also, by (vii) of 2.1, $(ab)^n \leq b^n a^n < b^{n+1}$. Thus b and ab form an anomalous pair, which is a contradiction. Similarly b and ba form an anomalous pair if $ba \leq ab$. The case when a and b are negative can be treated similarly.

The second statement is evident since by 3.22, S is o-isomorphic with a subsemigroup of positive real numbers under usual order and hence S has no anomalous pairs.

It is not necessarily true that if a cancellative t.o. semigroup is o-Archimedean, then it has no anomalous pairs. Let S be a commutative semigroup generated by a and b. Every element of S is of the form $a^n b^m$ with $n,m \geq 0$. Define $a^n b^m < a^k a^\ell$ if either $n + m < k + \ell$ or $n + m = k + \ell$ and $n < k$. S is o-Archimedean cancellative t.o. semigroup but a and b form an anomalous pair.

The basic step involved in the proof of Alimov's theorem about the characterization of subsemigroups of positive real numbers, is that the group of quotients of a t.o. semigroup, if exists, is o-Archimedean when the t.o. semigroup has no anomalous pairs. In general the group of quotients of a t.o. semigroup, if exists, need not be o-Archimedean. The group of quotients of the semigroup in the example cited in the above paragraph is not o-Archimedean. For $a > e$ and $ab^{-1} > e$, where e is the identity of the group of quotients. But $a^n b^{-n} < a$ for all n since $a^n < ab^n$.

THEOREM 4.2. *Let S be a commutative, cancellative positively t.o. monoid. Then S admits a group G of quotients and G is o-Archimedean iff S contains no anomalous pairs.*

PROOF. By 1.2, G exists. If G is o-Archimedean, then by Hölder's theorem, G can be embedded in the additive group of all real numbers and hence S has no anomalous pairs. To show the sufficiency, suppose that G is not o-Archimedean. Then there exist elements a,b,c in S such that $a > c$ and $a^n c^{-n} < b$ for every natural number n. Then bc and ba form an anomalous pair, which leads to the contradiction. For $c^n < a^n < bc^n$, whence

$$b^n a^n < b^n bc^n = b^{n+1} c^n \leq b^{n+1} c^{n+1}$$

$$b^n c^n < b^n a^n = (ba)^n \leq (ba)^{n+1}$$

THEOREM 4.3. *A positively t.o. semigroup S is o-isomorphic with a subsemigroup of the additive group of real numbers if, and only if, it satisfies the following conditions:*

(i) S contains no anomalous pairs;

(ii) S is cancellative.

PROOF. Clearly that subsemigroups of the additive group of real numbers satisfy (i) and (ii). Conversely let S satisfy those two conditions. By 4.1, S is o-Archimedean. We claim now that S is commutative. As observed in the proof of Theorem 4.1, every element is strictly positive ($xy > x$; $xy > y$ for every x and y) by cancellation. Suppose $ab < ba$ for some a and b in S. Then $(ba)^n < a(ba)^n < a(ba)^n b = (ab)^{n+1}$ and $(ab)^n \leq (ba)^n < (ba)^{n+1}$.

Hence ab and ba form an anomalous pair, which contradicts (i). Similarly it can be proved that $ba < ab$ is inadmissible. Thus S is commutative. By adjoining the identity 1 if S does not contain 1, we can treat S a cancellative commutative monoid and hence we can construct the group G of quotients of S. G can be totally ordered as in 1.2. Then by 4.2, G is o-Archimedean and hence G is embedded in the additive group of real numbers by Hölder's theorem.

Fuchs observed that it is possible to replace cancellative condition in the above theorem by a weaker one.

THEOREM 4.4. *A positively t.o. semigroup S is o-isomorphic with a subsemigroup of the additive group of real numbers iff S satisfies the following conditions:*

(i) S contains no anomalous pairs;

(ii) S is o-Archimedean;

(iii) S contains no maximal element unless it consists of a single element.

PROOF. It suffices to show that if S satisfies the three conditions then S is o-isomorphic with a subsemigroup of the additive group of real numbers, since the converse is trivial. By virtue of 4.3, we have to prove only that S is cancellative. Firstly we shall prove that $ab < b$ if $a,b \in S$ and $a \neq 1$ (if the identity 1 exists). By (iii) there exists an $c > b$ and by o-Archimedean property there exists a natural number n such that $a^n \geq c$. Clearly $ab \geq b$ by positive order. Now if $ab = b$, then $b = a^n b \geq a^n \geq c > b$, which is not true. Hence $ab > b$. We claim that S is commutative. Suppose $ab \neq ba$ for some a,b in S. Then a and b are different from identity. Then for every natural number n, $(ba)^n < a(ba)^n b = (ab)^{n+1}$ from the above, since $(ba)^n = 1$ implies $b = 1$ by (iii) of 2.1. Also $(ab)^n < b(ab)^n a = (ba)^{n+1}$ for every natural number n. This contradicts (i). Finally we assert that S is cancellative. Assume that $ab = ac$ and $b < c$. Clearly $(ab)^2 = (ac)b = (ab)c = ac^2$ and so by induction $ab^n = ac^n$ for every natural number n. By strict positive order (which is proved above), $ac^n < ac^{n+1} = ab^{n+1}$, which implies $c^n < b^{n+1}$ for every natural number n. Obviously $b^n \leq c^n < c^{n+1}$. This contradicts (i). Thus S is cancellative.

Other than the subsemigroups of additive group of real numbers, there exist examples of semigroups without anomalous pairs. Positively t.o. nilsemigroups have no anomalous pairs. Any t.o. chain with more than two elements contains no anomalous pairs. The semigroup in the first example is o-Archimedean and Archimedean whereas the second one is not o-Archimedean and also not Archimedean. We have noted above that cancellative t.o. semigroups without anomalous pairs are o-Archimedean but the converse need not be true. Now we shall determine conditions when positively t.o. semigroups contain no anomalous pairs.

THEOREM 4.5. *Let S be a positively t.o. semigroup. Then S has no anomalous pairs iff S is a semilattice of semigroups* S_α, *where* S_α *is a nilsemigroup or a subsemigroup of the additive group of real numbers without 0.*

PROOF. Suppose that S has no anomalous pairs. By 2.10, the positively t.o. semigroup S is a semilattice of semigroups S_α, where S_α is o-Archimedean semigroup and S_α does not contain identity if $|S_\alpha| > 1$. If $|S_\alpha| = 1$, S_α is trivially a nilsemigroup. Let $|S_\alpha| > 1$. Then S_α is a nilsemigroup if S_α contains a maximal element by 2.8. If S_α does not contain a maximal element, then S_α is a subsemigroup by the additive group of real numbers of 4.3. Conversely suppose now that S is a semilattice of semigroups S_α, where S_α is as described above. Each S_α can be treated as a subsemigroup of S. Let $a \in S_\alpha$ and $b \in S_\beta$. If S_α and S_β are nil, then we can choose a natural number n such that $a^n = e$ and $b^n = f$, where e and f are zeros of S_α and S_β respectively and so if a and b form an anomalous pair, then $e < f$ and $f < e$. This is absurd. If S_α and S_β

are subsemigroups of the additive group of real numbers, clearly a and b cannot form an anomalous pair. If S_α is nil and and if S_β is a subsemigroup of the additive group of real numbers, then $a^n = a^{n+1} = \ldots = e$, where e is the zero of S_α and so $a^n < b^{n+1}$ and $b^{n+1} < a^{n+2}$ would imply $e < b^{n+1} < e$, which again is not true. Thus S has no anomalous pairs.

THEOREM 4.6. *Let S be a positively t.o. semigroup without identity. Then S is an o-Archimedean semigroup containing no anomalous pairs iff S is nil or S is a subsemigroup of the additive group of real numbers.*

PROOF. Suppose that S is o-Archimedean without anomalous pairs. Then by 4.5, S is a semilattice of subsemigroups S_α where S_β is either nil or a subsemigroup of the additive group of real numbers. If one of S_α is nil, then the zero element of S_α is an idempotent in S and hence $S = S_\alpha$ is a nilsemigroup. If no S_α is nil, then every S_α is a subsemigroup of the additive group of real numbers and hence S is also a subsemigroup of the additive group of real numbers. The converse is evident.

THEOREM 4.7. *The following are equivalent for a t.o. semigroup S:*

(i) S is a positively ordered semigroup containing no anomalous pairs and $b \neq ba$ for every a,b in S.

(ii) S is a subsemigroup of the additive group of real numbers.

(iii) S is a left cancellative positively ordered semigroup containing no idempotents and no anomalous pairs.

PROOF. (i) $\Rightarrow$ (ii). Clearly S has no idempotents, and so in the representation of S given in Theorem 4.5, no S_α can be nil and thus (ii) is evident by 4.5. (ii) $\Rightarrow$ (iii) is trivial.

(iii) $\Rightarrow$ (i). If $b = ba$ for some a,b in S, then $ba = ba^2$ and hence $a = a^2$, which is not true.

The relation of a pair of elements being an anomalous pair is a congruence in some cases. This fact is utilized by Hion and Kowalski to prove that o-Archimedean positively t.o. semigroups can be homomorphically mapped into the additive group of real numbers.

LEMMA 4.8. *Let S be an o-Archimedean, cancellative positively t.o. semigroup. Let* ρ *be a relation on S defined by:* $a\rho b$ *if* $a = b$ *or a and b form an anomalous pair. Then* ρ *is congruence on S such that every* ρ*-class is convex.*

PROOF. Clearly ρ is reflexive and symmetric. Suppose $a\rho b$ and $b\rho c$. If $a = b$ or $b = c$ then trivially $a\rho c$. Now consider the case when $a^n < b^{n+1}$, $b^n < a^{n+1}$, $b^n < c^{n+1}$, and $c^n < b^{n+1}$ are satisfied at the same time for every n. If $a^{n+1} \leq c^n$ for some natural number n, then

$$a^{2n+2} \leq c^{2n} < b^{2n+1} < a^{2n+2}$$

which is a contradiction. Thus $c^n < a^{n+1}$ for every n. Similarly $a^n < c^{n+1}$ for every n. Thus $a\rho c$ and hence ρ is transitive. ρ is a congruence. For, let $a\rho b$ and let c be any arbitrary element of S. If $a = b$, then $ac = bc$ and $ca = cb$. So $ac\rho bc$. Now assume for definiteness $a < b$. Then $ac < bc$ by cancellation, which implies $(ac)^n < (bc)^n < (bc)^{n+1}$ since from $(bc)^n = (bc)^{n+1}$ we have $bc = (bc)^2$; $bc = 1$ and hence $b = c = 1$ by (iii) of 2.1. Hence $a < 1 \leq a$ by positive order, which is not true. Now by (viii) of 2.1 and (x) of 2.2, we have:

(1) if $ac \leq ca$ and $bc \leq cb$, then
$(bc)^n \leq c^n b^n < c^n a^{n+1} \leq (ac)^{n+1}$;

(2) if $ac \leq ca$ and $cb \leq bc$, then
$(bc)^n \leq b^n c^n < a^{n+1} c^n \leq (ac)^{n+1}$;

(3) if $ca \leq ac$ and $bc \leq cb$, then
$(bc)^n \leq c^n b^n < c^n a^{n+1} \leq c^{n+1} a^{n+1} \leq (ac)^{n+1}$;

(4) if $ca \le ac$ and $cb \le bc$, then

$$(bc)^n \le c^{n+1}b^n < c^{n+1}a^{n+1} \le (ac)^{n+1}$$

for every n. Thus $ac\rho bc$. Similarly $ca\rho cb$. Thus ρ is a congruence relation. We claim now each ρ-classes convex. For this, suppose $a < b < c$ with $a\rho c$. Then $a^n < c^{n+1}$ and $c^n < a^{n+1}$ for every natural number n. Therefore $c^n < a^{n+1} \le b^{n+1}$, $b^n \le c^n < a^{n+1} \le c^{n+1}$ for every n and so $c\rho b$ and thus a, b and c belong to the same ρ-class.

LEMMA 4.9. *Let S be an o-Archimedean positively t.o. semigroup containing no maximal element. Then the relation ρ as defined in 4.8 is a congruence relation.*

PROOF. As in the proof of 4.8, ρ is a convex equivalence relation. Let $a\rho b$ and c be any arbitrary element in S. If $a = b$, then clearly $ac = bc$ and $ca = cb$ and hence $ac\rho bc$ and $ca\rho cb$. Now suppose that (a,b) form an anomalous pair. For definiteness we may assume $a < b$ and $c \ne 1$. If $ac = 1$ or $bc = 1$, we have by 2.1, $c = 1$. Suppose $(ac)^m = (bc)^{m+1}$ for some integer $m > 0$. Then, since $a < b$, $(bc)^m \ge (ac)^m = (bc)^{m+1}$ and hence $(bc)^m = (bc)^{m+1}$, which implies $(bc)^m$ is an idempotent. Since the only idempotents in o-Archimedean positively t.o. semigroups are 1 and 0 (maximal element), the fact that $(bc)^m$ is an idempotent is not true. Hence $(ac)^m \ne (bc)^{m+1}$ for every integer $m > 0$. As proved in 4.8, we have, by considering all possible different cases, $(bc)^n \le (ac)^{n+1}$ and $(ac)^n \le (bc)^{n+1}$ for every natural number n. Note here we need not have strict inequalities as in cancellative case. But, however, as proved above, since $(ac)^m \ne (bc)^{m+1}$ for every natural number m, we have $(ac)^n < (bc)^{n+1}$. It suffices now to show that $(bc)^n \ne (ac)^{n+1}$ for every natural n by virtue of the above observation, in order to prove $ac\rho bc$. If $(bc)^n = (ac)^{n+1}$ for some $n > 0$, then $(bc)^{2n} = (ac)^{2n+2} \ge (ac)^{2n+1} \ge (bc)^{2n}$ and so $(ac)^{2n+1} = (ac)^{2n+2}$, which implies a contradiction as proved previously. Similarly we can show $ca\rho cb$. Thus ρ is a convex congruence relation.

THEOREM 4.10 [Hion]. *Let S be a cancellative, positively t.o. o-Archimedean semigroup. Then there exists an o-homomorphism of S into the additive semigroup of the nonnegative real numbers such that two distinct elements of S have the same image iff they form an anomalous pair.*

PROOF. The congruence ρ defined in 4.8 is a congruence on S such that every ρ-class is convex. Therefore $\bar{S} = S/\rho$ is a positively t.o. semigroup. $\bar{S}$ has no anomalous pairs since if $\bar{a}$ and $\bar{b}$ form an anomalous pair, then a and b form an anomalous pair in S, where $a \in \bar{a}$ and $b \in \bar{b}$. Clearly $\bar{S}$ is cancellative. Hence by 4.3, $\bar{S}$ is o-isomorphic with a subsemigroup of positive real numbers. Thus the canonical homomorphism $S \to \bar{S}$ is the required o-homomorphism.

Utilizing Lemma 4.9 and a similar proof of 4.10, one obtains:

THEOREM 4.11 [Kowalski]. *Let S be an o-Archimedean positively t.o. semigroup containing no maximal element. Then there exists an o-homomorphism of S into the additive semigroup of the nonnegative real numbers such that two distinct elements of S have the same image iff they form an anomalous pair.*

In 4.10 and 4.11, we have noted certain positively t.o. semigroups with o-Archimedean condition can be mapped into a subsemigroup of positive real numbers. Recall that in 2.22 and 3.7, we exhibited o-homomorphism mappings for positively t.o. semigroups not necessarily satisfying o-Archimedean condition.

PROBLEMS

1. Find the necessary and sufficient conditions for a positively t.o. semigroup to be embeddable in a right n.t.o. semigroup. Clifford provides an example that this is not the case always.

2. In the positively t.o. semigroup S, let A be the set of all o-Archimedean elements and Q be the union of all proper completely prime ideals. Find conditions when $Q \setminus A$ and $S \setminus Q$ are convex and when $Q \setminus A$ is o-Archimedean.

3. Describe positively t.o. semigroups in which every N-class is convex.

4. Let S be a ζ-indecomposable positively t.o. semigroup containing a central cancellable element and containing at least two elements. It is proved in 2.40, S is of the form $N \times G$. Under what conditions is the converse true?

5. Let N be the additive semigroup of nonnegative integers and G be any group. Let I be a function: $G \times G \to N$ satisfying:

(i) $I(\alpha,\beta) + I(\alpha\beta,\gamma) = I(\alpha,\beta\gamma) + I(\beta,\gamma)$ for every α,β,γ in G

(ii) $I(\varepsilon,\varepsilon) = I(\alpha,\varepsilon) = I(\varepsilon,\alpha) = 1$ for every α in G, where ε is the identity of G.

Define multiplication in $N \times G$ by:

$(m,\alpha)(n,\beta) = (m + n + I(\alpha,\beta),\alpha\beta)$. Find all possible positive orders on $N \times G$. What conditions must we impose on G and the I-function in order for $N \times G$ to be an o-Archimedean positively t.o. semigroup?

6. We have proved that finitely generated right n.t.o. semigroups are left n.t.o. if and only if they are commutative. Is commutativity the necessary and sufficient condition for a right n.t.o. semigroup to be a left n.t.o. semigroup always?

7. Prove 3.35 with only one-sided n.t.o. condition.

8. Characterize right n.t.o. nilsemigroups not containing minimal elements.

9. When ordinally irreducible complete positively t.o. semigroups are o-Archimedean? For n.t.o. semigroups, both conditions are equivalent. Hence the question arises whether o-Archimedean condition implies the condition of being complete for arbitrary positive t.o. semigroups and if not, when is it possible?

REFERENCES

1. M. Petrich, *Introduction to Semigroups*, Merrill Research and Lecture Series, Merrill Publishing Company, Columbus, Ohio (1973).
2. L. Füchs, *Partially Ordered Algebraic Systems*, Pergamon Press (1963).
3. T. Saito, Note on the Archimedean property in an ordered semigroup, *Proc. Japan. Acad.*, 46(1970), 64-65.
4. E. Ya. Jabovich, Fully ordered semigroups and their applications, *Russian Math. Surveys*, 31(1976), 147-216.
5. O. Kowalski, On Archimedean positively fully ordered semigroups, *Ann. Uni. Sci. Budapest Ser. Math.*, 8(1965), 97-99.
6. P. Conrad, Semigroup of real numbers, *Portugaliae Math.*, 18(1959), 199-201.
7. A. H. Clifford, Naturally totally ordered commutative semigroups, *Amer. Jour. Math.*, 76(1954), 631-646.
8. A. H. Clifford, Completion of semi-continuous ordered commutative semigroups, *Duke Math. Jour.*, 26(1959), 41-59.
9. V. S. Krishnan, L'extension d'une $(<,\cdot)$ algebra a une $(\Sigma^*,\cdot)$ algebre, I-II, *C. R. Acad. Sci. Paris*, 230(1950), 1447-1448, 1559-1561.

BIBLIOGRAPHY

- A. H. Clifford, Naturally totally ordered commutative semigroups, Amer. Jour. Math., 76(1954), 631-646.
- A. H. Clifford, Ordered commutative semigroups of the second kind, Proc. Amer. Math. Soc., 9(1958), 682-687.
- A. H. Clifford, Totally ordered commutative semigroups, Bull. Amer. Math. Soc., 64(1958), 305-316.
- A. H. Clifford, Completion of semi-continuous ordered commutative semigroups, Duke Math. Jour., 26(1959), 41-59.
- A. H. Clifford and G. B. Preston, The Algebraic Theory of Semigroups, Math. Surveys No. 7, Amer. Math. Soc., Vol. I(1961), Vol. II(1967).
- P. Conrad, Ordered semigroups, Nagoya Math. Jour., 16(1960), 51-64.
- P. Conrad, Semigroup of real numbers, Portugaliae Math., 18(1959), 199-201.
- N. E. Domoshnitskaya, The generalized measurement of elements of almost monotonically ordered semigroups, Nanchn. Trudy Permsk. Politekn. Inst., 7(1960), 41-56.
- L. Füchs, Partially Ordered Algebraic Systems, Pergamon Press (1963).
- L. Füchs, Note on fully ordered semigroups, Acta Math. Acad. Sci. Hung., 12(1961), 255-259.
- L. Füchs, On the ordering of quotient rings and quotient rings and quotient semigroups, Acta Sci. Math. Szeged., 22(1961), 42-45.
- O. Hölder, Die Axiome der Quantität und die Lehre vom Mass, Ber. Verh. Sachs. Ger. Wiss. Leipzig, Math. Phys. el, 53(1901), 1-64.

- E. V. Huntington, A complete set of postulates for the theory of absolute continuous magnitude, *Trans. Amer. Math. Soc.*, 3(1902), 264-279.
- E. Ya. Jabovich, Fully ordered semigroups and their applications, *Russian Math. Surveys*, 31(1976), 147-216.
- E. Ya. Jabovich, The orderability of some classes of groupoids, *Semigroup Forum*, 9(1974), 139-154.
- F. Klein-Barmen, Über gewisse Halbverbande und kommutative semigruppen, I-II, *Math. Zeitschrift*, 48(1942-43), 275-278, 715-734.
- O. Kowalski, On Archimedean positively fully ordered semigroups, *Ann. Uni. Sci. Budapest Ser. Math.*, 8(1965), 97-99.
- V. S. Krishnan, L'extension d'une $(<,\cdot)$ algebre a une $(\Sigma^*,\cdot)$ algebre, I-II, *C. R. Acad. Sci. Paris*, 230(1950), 1447-1448, 1559-1561.
- Mario Petrich, *Introduction to Semigroups*, Merrill Research and Lecture Series, Merrill Publishing Company, Columbus, Ohio (1973).
- T. Saito, The Archimedean property in an ordered semigroup, *Jour. Aust. Math. Soc.*, 8(1968), 547-556.
- T. Saito, Note on the Archimedean property in an ordered semigroup, *Proc. Japan Acad.*, 46(1970), 64-65.
- M. Satyanarayana and N. Srihari Nagore, On naturally totally ordered semigroups (submitted for publication).
- M. Satyanarayana, On naturally totally ordered semigroups II (submitted for publication).
- M. Satyanarayana, On characterizations of a class of naturally ordered semigroups, to appear in *Pacific Journal Math*.
- M. Satyanarayana, On positively totally ordered semigroups (submitted for publication).
- B. M. Schein, On certain classes of semigroups of binary relations (Russian), *Sibirsk. Matem. Zurnal*, 6(1965), 616-635.
- B. M. Schein, On the theory of generalized groups and generalized heaps, in *The Theory of Semigroups and Applications* I, Izad. Saratov Univ. Saratov (1965).
- T. Tamura, Commutative nonpotent Archimedean semigroups with cancellation law I, *Jour. Gakugei. Tokushima Univ.*, 8(1957), 5-11.

INDEX